COMPUTER SCIENCE, TECHNOLOGY AND APPLICATIONS

WIRELESS NETWORKS

Computer Science, Technology and Applications

Additional books in this series can be found on Nova's website under the Series tab.

Additional e-books in this series can be found on Nova's website under the eBooks tab.

COMPUTER SCIENCE, TECHNOLOGY AND APPLICATIONS

WIRELESS NETWORKS

S. ANANDAMURUGAN

AND

P. S. NANDHINI

nova publishers

New York

NOTICE TO THE READER

Library of Congress Cataloging-in-Publication Data

ISBN: 978-1-63485-178-7

Published by Nova Science Publishers, Inc. † New York

CONTENTS

Preface		**vii**
Chapter I	Introduction to Wireless Networks	**1**
Chapter II	MAC Protocols for ADHOC Networks	**45**
Chapter III	Overview of Basic Routing Protocols	**107**
Chapter IV	Overview of Advanced Routing Protocols	**143**
Chapter V	Energy Conservation Schemes in Wireless Networks	**179**
About the Authors		**201**
Index		**203**

PREFACE

This book gives a detailed introduction to Wireless Networks. It is very useful for those who are new to wireless. The aim of this book is to give maximum guidance to the students, faculty, research scholars and people working with computers. Suggestions for improvement will be appreciated and incorporated.

Dr. S. Anandamurugan
P. S. Nandhini

Chapter I

INTRODUCTION TO WIRELESS NETWORKS

OBJECTIVE

- *To learn about the fundamentals of Wireless Communication Technology*
- *To know the characteristics of Wireless Channels*
- *To understand the concepts and architectures of MANETs and WSNs*
- *To be familiar with the application of Adhoc and sensor networks*
- *To interpret the design challenges in Adhoc and sensor Networks*

1.0. FUNDAMENTALS

- Wireless communication is one of the most important areas in the field of communication.
- The two fundamental aspects of wireless communication are
  - Fading and
  - Interference.
- The phenomenon of *fading* refers to channel strength and it varies by time because of *small-scale effects* of multipath fading and *large-scale effects* such as path loss due to obstacles in the network.
- In the wired network, every transmitter-receiver pair has a point-to-point link.
- In wireless the users communicate through air and there is significant *interference* between them.

- This interference could be between transmitter and a common receiver, between single transmitter and multiple receivers.
- The traditional design of the wireless system is focused on increasing the reliability of air interface.
- Now-a-days the focus is on increasing *spectral efficiency.*

Wireless Communication

- The information which is transferred between two or more points that are not connected through links is called *wireless communication.*
- Wireless technology commonly uses radio waves for communication.
- The radio waves are used in *short-distance communication*, such as the television operating in a few meters and these radio waves are also used in deep-sea communication.
- Wireless communication generally works through *electromagnetic signals* that broadcast the message through air, physical environment or atmosphere; creating the wireless communication bridge between transmitter and receiver.
- Communication has various forms, technology and delivery methods such as,
    - Satellite communication
    - Mobile communication
    - Wireless network communication
    - Infrared communication
    - Bluetooth communication
- If a signal propagates through the wireless channel, the *random fluctuation* is experienced by the surrounding objects that move because of changing reflection and attenuation.
- The channel characteristics change randomly with time, which makes it difficult to design a reliable system with guaranteed performance.
- The most important area in wireless communication is *security* while implementing because in the wireless system the air waves are subject to snooping from anyone with an RF antenna.
- There is no security in the analog cellular system and one can easily listen to conservation by *scanning the frequency band.*
- But in the digital cellular system there is some level of *encryption*, and with enough knowledge, the encryption methods can be cracked.

- The design of wired network is based on the layered approach.
- In *wireless network,* the layer includes the following:
  - *Link or physical layer* - To transmit the bits over a communication medium.
  - *Access layer* – To Handle shared access to the communication medium
  - *Network and transport layer* - To route the data across the network.
  - *Application layer*- To Handle the end-to-end data rates and delay constraints of the application.
- The *layering methodology* reduces complexity and facilitates modularity and standardization which leads to inefficiency in performance loss because of lack in global design optimization.
- But in wired network there is large capacity and good reliability.
- Wireless link may lead to poor performance due to network topology changes over time.
- The wireless systems that operate today are:
  - Cellular telephone system
  - Cordless phones
  - Wireless LANs
  - Wide area wireless Data services
  - Broadband wireless access
  - Paging system
  - Satellite networks
  - Low-cost Low-power radio: Bluetooth and Zigbee
  - Ultra wideband radios

1.1. Electromagnetic Spectrum

- The principle of wireless communication is based on the broadcast and reception of electromagnetic waves.
- These waves are characterized by their *frequency* (f) or their *wavelength* (λ).
- Frequency is the *number of cycles per second* for the wave and it is measured in *Hertz* (Hz).
- *Wavelength* is the distance between two consecutive maxima or minima in the wave.

- The propagation speed of the waves varies from *medium to medium*, except in a vacuum where electromagnetic waves travel at the same speed, the *speed of light.*

$$C = \lambda \times f$$

Where c = *speed of light* (3×10^{8}m/s)
f = *frequency of the wave* in Hz
λ = *wavelength* in meter.

- The *low-frequency bands* are radio, microwave, infrared and visible light.
- This portion of the spectrum is used for information transmission by modulating amplitude, frequency or the phase.
- The *high-frequency waves* are x-rays and gamma rays.
- The propagation of information through high-frequency wave appears better in theory but in practice it is difficult to generate these waves.
- The high frequency waves do not propagate through buildings.
- The amount of information carried by electromagnetic waves is determined by the width of the wavelength band.
- Radio waves are generated easily and they can facilitate communication both indoor and outdoor
- They have the ability to travel long distances.
- The characteristics of the transmission are determined by the frequency of radio waves.
- At low frequency the radio waves can pass through obstacles easily.
- The high frequency waves are prone to absorption by rain and reflected by the obstacles.
- The propagation waves of VLF, LF and MF are also called *ground waves*.
- The maximum transmission ranges of these waves are few hundred kilometers.
- These are used for low bandwidth transmission such as Amplitude Modulated (AM) radio broadcasting.
- The HF and VHF are absorbed by the atmosphere near the earth's surface.
- The *sky wave* is a portion of radiation that radiates outwards and upwards on to the ionosphere.

- The *ionosphere* contains the ionized particles, and these ionized particles reflect the sky wave to the earth.
- The sky waves are used in military communication and ham radio operations.
- Microwaves travel in straight line and are transmitted narrowly.
- These waves were used for long-distance telephony, before they were replaced by fiber optics and used also in mobile phones and television transmission.
- The transmission has a high signal-to-noise ratio.
- *Signal to Noise Ratio* (SNR) is defined as the ratio between signal powers to noise power on a transmission medium.
- *Infrared waves* are used for short-range communication and they are used in television, VCR and stereo remote control.
- The spectrum contains *visible light* just after the infrared portion.
- Unguided optical signaling uses the visible light and provides very high bandwidth at low cost.
- Lasers are used in the light-wave transmission to connect LAN on two building antennas.
- The main disadvantage in this is that it is very difficult to focus on a narrow uni-directional laser beam and also these waves cannot penetrate through rain or thick fog.

1.2. Radio Propagation Mechanism

- Radio signals are affected by the objects on the path or media through which they travel. The path by which the radio signal propagates controls the level and quality of signal it receives.
- Ray-optical method can be used to describe radio propagation within and even outside buildings. Radio signal propagation is important for designing or operating a radio system.

Radio propagation with ray optics has three basic mechanisms,

- Reflection and Transmission
- Diffraction
- Scattering

1.2.1. Reflection and Transmission

- When a radio wave falls on any other medium having different electrical properties, a part of it is transmitted into it, while the rest are reflected back.
- The amount of energy that is reflected depends upon the polarization of the wave.
- When the radio wave hits an object on the ground, the wave gets reflected by that object.
- Reflection causes a phase of 180 degree between the incident and reflected rays.
- For example, if a ray incident happens on the ground, walls of buildings, the ceiling and the floor undergo reflection and transmission.

1.2.2. Diffraction

- When a ray hits the edge of an object, the ray bends at the edges of the object thereby propagating in different directions.
- The dimensions of the object causing the diffraction are comparable with the wavelength of the ray being diffracted.
- Because of the bending, the ray reaches places behind the object and it cannot reach the line of sight transmission.

1.2.3. Scattering

- Scattering occurs when the ray travels through a medium, which contains many objects whose dimension is small when compared to the wavelength.
- They get scattered into several weaker outgoing signals.

1.3. Characteristics of the Wireless Channel

- The Wireless Channel has a variety of transmission impediments such as path loss, interference, and blockage.

- Due to these factors the range, data rate and the reliability of the wireless transmission is restricted.
- The wireless channel is characterized based on the following:
  - Path Loss
  - Fading
  - Interference
  - Doppler Shift

1.3.1. Path Loss

- *The ratio of the power of the transmitted signal to the power of the same signal received by the receiver on the given path is said to be path loss.*
- For designing and deploying a wireless communication network, the estimation of the path loss is important.
- Path loss is dependent on the radio frequency and nature of the terrain.
- In designing a network, several models are required to describe the variety of transmission environments.
- The *free space propagation* is the simplest path loss model i.e., there is a direct path signal between transmitter and receiver, with no atmospheric attenuation or multipath components.
- The relation between the transmitted power P_t and received power P_r is given by,

$$\mathrm{Pr} = P_t G_t G_r \left(\frac{\lambda}{4\prod \mathrm{d}}\right)^2$$

Where,

G_t = transmitter antenna gains

G_r = receiver antenna gains, in the direction from the transmitter to the receiver

d = distance between the transmitter and receiver and

$\lambda = c/f$ = wavelength of the signal.

- Another path loss model is the *two-ray model or the two-path model.*
- In the free space model there is only one single path between the transmitter and the receiver.

- But in the two-ray model, the signal reaches the receiver through *multiple paths*.
- This model assumes that the signal reaches the receiver through two paths, one a *line of sight path* and the other is the path through which the reflected wave is received.
- According to the two-ray model, the path loss is given by

$$\mathrm{Pr} = P_t G_t G_r (\frac{h_t h_r}{\mathrm{d}^2})^2$$

Where,
P_t = transmitted power
G_t = transmitter antenna gains
G_r = receiver antenna gains
d = the distance between the transmitter and receiver
h_r = height of the receiver
h_t= height of the transmitter

- In the general case, for isotropic antennas the received power is given by,

$$\mathrm{Pr} = P_t G_t G_r (\frac{\lambda}{4\prod})^2 \frac{1}{\mathrm{d}^\gamma}$$

Where,
γ = propagation coefficient that varies between 2 and 5.

1.3.2. Fading

The fluctuations in signal strength when received at the receiver end is referred to as fading and it is classified into,

- Fast fading/small-scale fading
- Slow fading/large-scale fading

Fast Fading

- Fast fading refers to *the rapid fluctuations in the amplitude, phase or multipath delays of the received signal due to interference* between multiple versions of the same transmitted signal arriving at the receiver at different times.
- *Delay spread* is expressed as the time between the reception of the version of the signal and the last enclosed signal.
- The cause of fast fading is the multipath propagation of the transmitted signal, because of the three propagation mechanisms namely reflection, diffraction, and scattering.
- A variation in the power level of the received signal because of multiple signal paths may sometimes add constructively or sometimes destructively at the receiver.
- The received signal of a fast fading signal follows a *Rayleigh distribution*, if there is no line of sight between the transmitter and receiver and Ricean distribution is one such path available.

Slow Fading

- *When objects partially absorb the transmission laying between the transmitter and receiver* it is called slow fading.
- This also occurs when the receiver is inside a building and the radio wave must pass through the walls of the building or when a receiver is temporarily shielded from the transmitter by a building.
- Slow fading is also referred to as *shadow fading* because the object that cause the fade, which may be a large building or other structure, blocks the direct transmission path from the transmitter to the receiver.
- Diversity mechanism is based on the independent path between the same transmitter and receiver nodes experiencing the independent fading effects.
- The independent paths can be distinct in space, time, frequency and polarization.
- The aim of the *time diversity mechanism* is to spread the data over time so that the effects of burst errors are minimized.

- *Frequency diversity mechanism* spreads the transmission over a wider spectrum or uses multiple carriers for transmitting the information.
- *Space diversity* involves the uses of different physical transmission paths.

1.3.3. Interference

- *Interference* occurs in a wireless transmission from a wide variety of sources.
- The main form of interference is
    - Adjacent channel interference.
    - Co-channel interference.
    - Inter-symbol interference
- In *Adjacent channel interference*, the signals in nearby frequencies have components outside their allocated range and these components may interfere with on-going transmission in the adjacent frequencies.
- This interference can be avoided fully by introducing *guard bands* between the allocated frequency ranges.
- Co-channel interference is sometimes referred to as *narrow-band interference*.
- This is due to other nearby systems using the same transmission frequency.
- In narrow-band interference, *frequency reuse* in cellular system can be minimized with the use of directional antennas and dynamic channel allocation methods.
- Another type of interference is *inter-symbol interference in which* distortion in the received signal is caused by the temporal spreading and consequent overlapping of individual pulse in the signal.
- If the temporal spreading of individual pulse goes beyond certain limit, the receiver becomes unable to reliably distinguish between change of state in the signal i.e., the bit pattern interpreted by receiver is not same as that sent by the sender.
- The technique used for combating inter-symbol interference is *adaptive equalization*.
- The mechanism involved in the adaptive equalization is to *gather the dispersed symbol energy into its original time interval.*

- For the equalization process, the *complex digital processing algorithm* is used.
- The main principle of adaptive equalization is the estimation of the channel pulse response to periodically transmitted well-known bit patterns, known as *training sequences.*

1.3.4. Doppler Shift

- Doppler shift is defined as the *change/shift in the frequency of the received signal when the transmitter and receiver are mobile with respect to each other.*
- If the transmitter and receiver move towards each other, the frequency of the received signal is higher than the transmitted signal.
- If they are moving away from each other, the frequency of the signal at the receiver will be lower than that of the transmitter.
- The Doppler shift f_d is given by,

$$f_d = \nu/\lambda$$

Where,
ν = Relative velocity between transmitter and receiver and
λ = Wavelength of the signal

1.4. Transmission Rate Constraints

Two important constraints are used to determine the maximum rate of the data transmission on a channel

- Nyquist's theorem
- Shannon's theorem.

1.4.1. Nyquist's Theorem

- The transmitted signal denotes the number of times per second the signal changes its value.

- The number of changes per second is measured in terms of *baud*.
- The baud rate is not same as the bit rate/data rate of the signal since each signal value may be used to convey multiple bits.
- This theorem gives the maximum data rate on a channel as

$$C= 2*b*\log_2 L$$

Where,
B = Bandwidth of the channel
L = Number of discrete signal level/voltage value
C = Maximum channel capacity.

1.4.2. Shannon's Theorem

- It is the ratio of signal power (S) to noise power (N) specified in decibels, i.e., $SNR= 10 \log_{10}(S/N)$.
- The Shannon theorem gives *maximum data rate in the noisy channel.*
- The maximum data rate C is given by,

$$C = B*\log_2 (1+(S/N))$$

Where, B = Bandwidth of the channel in Hz.

1.5. Concepts and Architectures

1.5.1. MANETs

- MANET is a collection of two or more wireless devices that have the capability to communicate with each other *without any centralized administrator.*
- Every node in a MANET acts as a host and a router.
- The network topology is *dynamic* due to the mobility of nodes, node departures and new node arrivals.
- Hence, an efficient routing protocol is needed to communicate with nodes.

Architecture

- In MANET the nodes are classified into
  - *Client or small mobile host* (SMH) i.e., it has limited processing, storage, communication and power resources.
  - *Server or large mobile host* (LMH) i.e., with large resources.
- Server has larger capacity of database management system to broadcast the data in order to satisfy the user queries.
- MANET is a collection of nodes in that every node has a *transmission range*.
- In that transmission range every node communicates with each other.
- A large mobile host initially has large transmission range and has a more powerful battery.
- If the power available to broadcast is reduced, then the transmission ranges of the node shrinks.
- The node in the network can operate in three modes that are designed to reduce the power used:
  - Active mode
  - Receive mode
  - Sleep mode

Active Mode

- In this mode, *the transmission and reception of messages* take place.
- Most of the power is used in this mode.

Receive Mode

- In this mode the CPU process the information and receives the message from other nodes and listens to the broadcast.

Sleep Mode

- The CPU does not receive or send the message to other nodes and it *does not process any information* during sleep mode.
- In this mode, the node is inactive and the node turns off itself for a short period of time without requiring power-up or re-initialization.

- When the node has no remaining power or the node is off, that node is not currently part of the network.
- This node cannot be reachable by other nodes and the node is disconnected from the entire network.
- When moving back to the range the node is re-connected.
- Network connectivity in MANET is provided in two ways,
    - Hierarchical network architecture
    - Flat-routed architecture

Hierarchical Network Architecture

- In this architecture, the whole network is partitioned into sub-networks.
- Each network by itself dynamically elects a node and that node acts as a *gateway* to other sub-networks.
- This sub-network builds a hierarchy that can be one-tier or multiple tier.
- The advantage of this network architecture is
    - Easy mobility management procedures
    - Better manageability

Flat-Routed Architecture

- In this architecture, *all nodes are identical* in terms of their responsibility and there is no gateway in this approach.
- The advantages of this architecture are
    - Optimal routing
    - Increased reliability because there is no single point failure and there are alternate routes in the network
    - Better load balancing
    - Reduced use of wireless resources

1.5.2. Wireless Sensor Networks

- Sensor networks consist of small, lightweight wireless nodes deployed in a distributed fashion to *monitor the environment or the system to measure physical parameters* such as temperature, pressure or relative humidity.

- In a sensor network each sensor node consists of three subsystems:
  - Sensor subsystem - Senses the environment,
  - Processing subsystem - Performs local computation on the sensed data
  - Communication subsystem - Responsible for message exchange with neighboring sensor nodes.
  - The sensor nodes have a limited sensing region, processing power, and energy.

Sensor Network Architecture

- The sensor networks are designed by factors such as scalability, fault tolerance and power consumption.
- The two basic kinds of sensor network architecture are
  - Layered architecture and
  - Clustered architecture

Layered Architecture

- Layer architecture has a *single base station* (BS), the *nearby nodes* have same hop-count to BS and there are layers of sensor nodes around the BS.
- Layered architecture is used in in-building wireless backbone, military sensor based infrastructure like multi-hop infrastructure network architecture.
- In the *in-building scenario*, for a wired network the BS acts as an *access point*, the small wireless nodes form a network to provide the wireless connectivity.
- *Hand-held devices* such as PDA are employed by the users to communicate via small nodes to the BS.
- In a *military operation*, the BS is a data-gathering and processing entity with a communication link.
- The *advantage* of layered architecture is that, every node provides a short-distance, low-power transmission to the neighboring layer nodes.

Unified Network Protocol Framework (UNPF)

- In the UNPF, the complete implementation of layered architecture for the sensor networks has a set of protocols.
- UNPF integrates three operations in its protocol structure:
    - Network Initialization and Maintenance Protocol
    - MAC Protocol
    - Routing Protocol

Network Initialization and Maintenance Protocol

- Sensor nodes are organized into different layers using the *broadcast capability* of the BS.
- A common control channel for all sensor nodes can reach the BS in one-hop.
- With a known *CDMA code* the BS broadcasts its identifier (ID) in a common channel.
- BS ID is recorded in all nodes that hear the broadcast.
- Then they send *beacon signal* with its ID to the BS at low power level.
- The nodes that have a single-hop distance from the BS form *layer one*.
- Then BS broadcasts the control packet with layer one IDs to all other nodes.
- Again all nodes send a beacon signal.
- The layer one nodes record the ID that they hear and this form a *layer two* because it is one-hop away from the layer one nodes.
- The layer one nodes inform the BS about layer two nodes by sending beacon signals which is then broadcast to the entire network.
- In this way the layered structure is built by sending beacons and BS broadcast.
- It also periodically updates the neighbor information and alters the layer architecture i.e., if a node is dead or moved out of the range by sending beacon signals.

MAC Protocol

- Network initialization is carried out in a *common control channel.*
- MAC protocol is used in the data transmission phase and it has *distributed TDMA receiver oriented channel* (DTROC).
- The BS assigns a reception channel to each node and the channel reuse avoids collision.
- The nodes schedule transmission slots for all neighbors and broadcast the schedule.
- The nodes can turn off when there is no send/receive operation.
- This saves energy and provides collision free transmission.
- The DTROC has two steps:
    - Channel allocation
    - Channel scheduling.
- In channel allocation algorithm the DTROC avoids hidden and exposed terminal problems.

Routing Protocol

- The BS can directly broadcast on the control channel.
- In layered architecture, the multi-hop data forwarding is from sensor nodes to BS.
- The node that forwarded the packet is selected based on the *remaining energy of the node.*
- This achieves a higher network life time.
- UNPF protocol modification is termed as UNPF-R.
- The transmission range of sensor nodes varies adaptively so that network performance is optimized.
- When the sensor nodes have a small transmission range they cause *network partitioning* and if there is a very large transmission range then it will reduce the *spatial reuse of frequencies.*
- Through an algorithm the optimized range is determined.
- The *centralized control algorithm*, the objective function is periodically evaluated by the BS.
- For the transmission range R, objective function $f(R) = \frac{\epsilon * d}{n/N}$

Where,
N = Total number of sensors in the system
n = Number of nodes in layer one
d = Average packet delay
ε = Energy consumption per packet.

- The new transmission range R\` is selected by the BS through the following ways,
  - If any sensor node does not receive a packet from the BS for some time interval, the transmission range is increased by Δr with the probability of [1-0.5 * (n/N)].
  - Otherwise, the transmission range is decreased by Δr with the probability of 0.5 * (n/N).
  - The re-evaluated objective function for new transmission range i.e., f(R\`) < f(R) is adopted.
  - Otherwise, R is modified to R\` with probability $e^{(f(R) - f(R`)) * (n/N)/T}$.

Clustered Architecture

- The sensor nodes are organized into clusters, each cluster having *one cluster-head.*
- In each cluster, the nodes can exchange the message with the cluster-head, and the head sends the message to the BS which acts as a access point in the wired network.
- Clustering can be extended to greater depth hierarchically.
- Clustered architecture is useful for sensor networks because of its inherent suitability for *data fusion.*
- In a cluster, all members can send data to the cluster-head, in cluster-head the data is gathered and the gathered information can communicate with the BS.
- The sensor networks should be self-organizing because the cluster formation and election of cluster-head is an *autonomous and distributed process.*
- The clusters are achieved through *network layer protocol.*

Low-Energy Adaptive Clustering Hierarchy (LEACH)

- LEACH is a *cluster-based protocol* that minimizes energy dissipation in sensor networks.

- The cluster-heads are randomly selected by LEACH which also performs periodic reelection because of high-energy dissipation in cluster-heads when communicating with BS.
- In an iterative manner the cluster-head is selected and each iteration is called *round.*
- The operation of LEACH is divided into two phases
  - Set-up phase
  - Steady phase
- In the *set-up phase*, the sensor nodes randomly choose the number between 0 and 1.
- If this is lower than the threshold for a node n, T(n), then the node becomes a cluster-head the threshold T(n) is calculated by,

$$T(n) = \begin{Bmatrix} \dfrac{P}{1 - P\left\{rxmod\left(\frac{1}{P}\right)\right\}} & if\ n \in G \\ 0 & otherwise \end{Bmatrix}$$

- Where,
  P = Desired percentage of nodes which are cluster-heads
  r = Current round
  G = Set of a node that has not been a cluster-head.
- All nodes spend *equal energy* in the set-up phase.
- All sensor nodes choose the nearest cluster-head to receive the advertisement based on received signal strength.
- Then the cluster-head assigns a *TDMA schedule* for their cluster members.
- The overhead of cluster formation is minimized in a steady phase.
- During the steady phase, based on the TDMA schedules, the data transmission takes place.
- The cluster-head performs the local computation i.e., *data fusion/aggregation.*
- The BS receives the gathered information from the cluster-head which leads to *energy conservation.*
- After a certain time period the cluster-head is selected by set-up phase.

1.6. Applications of Ad hoc and Sensor Networks

1.6.1. Application of Adhoc Networks

- The deployment of the Adhoc network is economically less demanding.
- It is deployed in many applications.
- Some of the applications are,
  - Military applications
  - Collaborative and Distributed Computing
  - Emergency Operation
  - Wireless Mesh Networks
  - Wireless Sensor Networks
  - Hybrid Network Architectures

Military Applications

- Adhoc networks are useful in *establishing communication* among a group of soldiers in tactical operations.
- Setting up the fixed infrastructure may not possible for communicating among group of soldiers, so the Adhoc wireless network is used.
- This application enables *quick and reliable communication.*
- For example if a leader wants to give orders to a group of soldiers or set of persons involved in an operation, this application is used.
- The routing protocol used must provide quick, secure, and reliable communication.
- *Secure communication* is required for military applications at any cost.
- For efficient communication and coordination services such as location tracking or other satellite-based services are used.

Collaborative and Distributed Computing

- A temporary communication infrastructure requires *quick communication* and *minimal configuration* through Adhoc wireless network.

- For example, if a group of researchers wants to share the files or presentation material during a conference, the formation of Adhoc wireless network with *reliable multicast routing* can be used.
- The distributed file sharing applications do not require the level of security expected in a military environment.

- But the reliability of data transfer is highly important as for distributing a file to other nodes in the network.
- The main goal of transmission is that all desired receiver nodes have the *replica of the transmitted file*.

Emergency Operation

- The major task of Adhoc wireless network is
    - Self-configuration of the system with minimal overheads
    - Being Independent of fixed or centralized infrastructure
    - To match the unavailability of conventional communication infrastructure.
- In a conventional communication infrastructure environment, the facilities could be destroyed due to war or natural calamities such as earthquake.
- Natural calamities are unexpected and affect a large number of people.
- So Adhoc wireless networks should be distributed and scalable to a large number of nodes.

Wireless Mesh Networks

- Wireless mesh networks are formed to provide an *alternate communication infrastructure* for mobile or fixed nodes without spectrum reuse constraints.
- The mesh topology provides an alternate path for data transfer between source and destination.
- This network also provides the most *economical data transfer capability*.
- The deployment scenario of wireless mesh networks includes,
    - Residential zones
    - Highways

- Business Zones
- Important civilian areas
- University campuses
- The main *advantage of mesh networks* are support for a
  - High data rate
  - Quick and low cost of deployment
  - Enhanced service
  - High scalability
  - Easy extendibility
  - High availability
  - Low cost per bit
- Mesh networks support large number of nodes.
- The deployment time of the network is much less compared to infrastructure based network.

Wireless Sensor Networks

- Sensor networks are used to provide wireless communication infrastructure among the sensors deployed in specific applications.
- Sensor node is a *tiny device* that is used for
  - Sensing the physical parameters
  - Processing the data gathered and
  - Communication over the network to the monitoring station.
- Sensor networks consist of numerous sensor nodes that are deployed in a particular region.
- The sensing activity can be *periodic or irregular.*
- An example for periodic type of sensing is temperature, humidity, and nuclear radiation.
- The following issues make sensor networks different from the Adhoc networks.

Mobility of Nodes

- Mobility of nodes is not a compulsory requirement in sensor networks.
- Sensor networks need not be designed to support mobility of sensor nodes.

Size of Network

- The nodes in the sensor network are larger than in Adhoc wireless networks.

Density of Deployment

- The density of nodes in sensor networks is different for different domains of application.

Power Constraints

- Sensor networks are power constraint than those in Adhoc wireless networks.
- This is because the sensor nodes are expected to operate in geographical conditions with the *minimum human resources and maintenance*.
- In such cases, the recharging of energy in sensor nodes is impossible.
- Running such a network with limited battery demands for efficient protocol at network, data link, and physical layer.
- The power source used in sensor networks can be divided into,
  - Replenishable power source
  - Non-replenishable power source
  - Regenerative power source

Replenishable Power Source

- In some application of sensor networks, the *power source is replaced* when the source is fully drained.

Non-Replenishable Power Source

- In some specific application, the power source *cannot be replenished* once the network has been deployed.
- *Sensor node replacement* is the only solution.

Regenerative Power Source

- Power sources employed in sensor networks could have a capability of a *regenerative power from physical parameters under measurement.*
- For example, if sensor nodes are used for sensing temperature at a power plant then it can use the power source that can generate power by using transducers.

Data/Information Fusion

- The aggregation of multiple packets into one before relaying it is referred to as *data fusion.*
- The main aim is *reducing bandwidth* consumed by redundant header packets and reducing the media access involved in transmitting multiple packets.
- The aim of information fusion is processing the sensed data at the intermediate nodes and relaying the outcome to the monitor node.

Traffic Distribution

- The communication traffic pattern varies for different domain applications.
- For example, in the *environment sensing application*, the packets are generated periodically indicating the status of the environmental parameter to the central monitoring station.
- This kind of traffic demands low bandwidth.
- If a sensor network used in military application, generating traffic on detecting certain events, these events might have time constraints for delivery.

Hybrid Wireless Networks

- Hybrid wireless network is one of the major application areas of Adhoc wireless networks.
- Hybrid wireless architecture is a combination of *multi-hop cellular network* (MNC) and *integrated cellular Adhoc relay* (iCAR).
- The main concept of cellular network is *geographical channel reuse.*

- *Cell sectoring, cell resizing, and multi-tier cell* are techniques that have been used to increase the capacity of cellular networks.
- It can also be increased by incorporating the properties of multi-hop relaying.
- In the above figure, when two nodes want to communicate with each other in the same cell, the connection is routed through the multiple wireless hops over the intermediate nodes.
- For efficient routing the base station maintains the information about the topology of the network.
- If node A wants to communicate with node B and if all nodes are capable of operating in MNC node, node A can reach node B directly as node B is in the transmission range of node A. When node C wants to communicate with node E where both are in the same cell, node C can reach node E through node D, which is the intermediate relay node.
- The major advantage of hybrid wireless network is,
    - High capacity than cellular network because of channel reuse
    - Flexibility and reliability is increased in routing by the use of multi-hop path
    - Better coverage and connectivity

1.6.2. Application of Wireless Sensor Networks

- A sensor network is used to perform a set of *high-level information processing* tasks such as detection, tracking or classification.
- These tasks are measured and the performance is well defined, including the detection of false alarms, classification errors, and tracking quality.
- Based on the application requirements, the application of sensor networks vary significantly.
- Sample commercial and military applications are
    - Environmental monitoring (e.g., traffic, habitat, security)
    - Industrial sensing and diagnostics (e.g., appliance, factory, supply chains)
    - Infrastructure protection (e.g., power grids, water distribution)
    - Battlefield awareness (e.g., multi-target tracking)

- Context-aware computing (e.g., intelligent home, responsive environment)

Sensor Network Application in Home Control

- Home control applications provide control, conservation and safety
- Sensor application in homes provides flexible management of *water, gas and electric utility*
- Sensing applications facilitate *flexible management* of lighting, heating and cooling systems
- Sensing applications provide the reception of *automatic notification* upon detection of unusual events
- Sensing applications are used for *security purposes* in homes

Sensor Network Application in Industrial Automation

- Sensing application helps to identify inefficiently operating equipments
- Sensor applications deployed in industries contribute to safety
- They reduce energy cost through optimized manufacturing processes
- To reduce human intervention, sensors automate the data acquisition from the remote sensors
- Sensors deployed in industry act as a *preventive maintenance factor*
- In Industry, sensors are deployed in
    - Lighting control
    - Gas, water and electric meters
    - Smoke detectors
    - Security services
    - Refrigerators

Sensor Network Application in Medical Applications

- Sensor networks are deployed in emergency cases, disaster response and stroke patient rehabilitation
- They provide home monitoring for elderly patients, chronic disease affected patients and physically disabled patients so as to reduce the *length of hospital stay*

- They help the physicians to study the effects of medical intervention programs

Sensor Network Application in Military Applications

- Military sensor networks are used to *detect and gain information* about enemy movement, explosions, etc.
- It is used to reduce staffing costs
- WSN is deployed through space vehicles to perform a variety of missions.
- WSN can replace single high-cost sensor assets with large arrays of distributed sensors for both security and surveillance applications.
- The added feature of robust, self-organizing networking makes WSN deployable by untrained troops in critical situations also
- Military applications require support for tactical and surveillance arrangements that employ reconfigurable sensors

Nanoscopic Sensor Applications

- WSN is used for biological sensing
- *Mobile robot* acting as a gateway in the communication between the robot and the sensor network.
- Nanoscopic sensors are deployed for preventive maintenance for equipment in a semiconductor manufacturing lab, etc.
- It is very useful for attractive for *earthquake-active zones.*

1.7. DESIGN CHALLENGES

1.7.1. Design Challenges in Wireless Sensor Networks

- A sensor network has a set of resource constraints such as *limited battery power and communication bandwidth.*
- Each sensor node operates on a *micro processor* and a small amount of memory for signal processing and task scheduling.
- It is also equipped with one or more sensing device such as *acoustic microphone arrays, video or still cameras, infrared, seismic or magnetic sensors.*

- Each sensor node communicates wirelessly when there are other nodes within its radio communication range.
- *Gateways* are used to route the user queries or command to appropriate nodes in a sensor network.
- It also routes sensor data to users who have requested or expected to utilize the information.
- Gateways have the *data repository* or storage device in addition to data logging at each sensor.
- The *repository* may serve as an intermediate between users and senders.
- Additionally, one or more storage device may be attached to the IP network, to collect the sensor data from a number of sensor networks and to support user-initiated browsing and search functions.
- The challenges that are faced while designing a sensor network system are:
    - Limited hardware
    - Limited support for networking
    - Limited support for software development

Limited Hardware

- Each sensor node has limited processing, storage and communication capabilities and limited energy supply.

Limited Support for Networking

- Each sensor node acts as a *router* and as an *application host.*
- The network is peer-peer with *mesh topology* and dynamic, mobile, and unreliable connectivity.
- There is no *universal protocol or central registry services* in sensor networking.

Limited Support for Software Development

- Sensor networking involves *dynamic collaboration* among the nodes; the tasks are typically real-time and massively distributed and must be capable of handling multiple competing events.

- Through local instruction the global properties are specified.
- For coupling between applications and system layers, the *software architecture* must be co-designed with information processing architecture.

1.7.2. Design Challenges in Wireless Adhoc Networks

- There are challenges that need to be considered while designing Adhoc networks.
- Deployment considerations like *installation, operation and maintenance* of Adhoc networks.
- The design, deployment, and performance of Adhoc network is affected by the following major issues:
    - Medium access scheme
    - Routing
    - Multicasting
    - Transport layer protocol
    - Pricing scheme
    - Quality of service provisioning
    - Self-organization
    - Security
    - Energy management
    - Addressing and service discovery
    - Scalability
    - Deployment considerations

Medium Access Scheme

- The medium access control (MAC) protocol is used in Adhoc networking to transmit the packet in the shared channel. The issues for designing a MAC protocol for Adhoc networks are as follows
    - Distributed operation
    - Synchronization
    - Hidden terminal
    - Exposed terminal

- Throughput
- Access Delay
- Fairness
- Real time traffic support
- Resource reservation
- Ability to measure resource availability
- Capability for power control
- Adaptive rate control
- Use of directional antenna
- Routing
- Mobility
- Bandwidth constraint
- Error-prone and shared channel
- Location-dependent contention
- Other resource constraints

Distributed Operation

- The Adhoc networks need to be operative in an environment with no centralized coordination.
- The design should be fully distributed for minimum control overhead.
- In case of partial coordination, *Polling based MAC protocol* is used.

Synchronization

- For TDMA-based system, *synchronization* is compulsory for management of transmission and reception of slots.
- The involvement of the system is based on the usage of scarce resources such as bandwidth and battery power.
- In synchronization, collision is increased by using the *control packets*.

Hidden Terminal

- The nodes are *hidden from the sender but they are reachable to the receiver*.
- In this case the hidden terminal causes collisions at receiver node.
- Hidden terminal significantly *reduces the throughput* in the network.

Exposed Terminal

- Exposed terminal refers to the nodes that are in the transmission range of the sender.
- To improve the efficiency of MAC protocol the node should transmit the packet in a controlled fashion without causing *collision.*

Throughput

- The MAC protocol is used in Adhoc network to *maximize the throughput.*
- In throughput enhancement, the channel utilization should be maximum while the collision and control overhead should be minimum.

Access Delay

- The MAC protocol should minimize the access delay and it is referred as the *average delay* that any packet experience to get transmitted

Fairness

- Fairness can be done either by the *node-based or flow-based approach.*
- The MAC protocol is used to provide equal share or weighted share of bandwidth to all nodes.

Real-Time Traffic Support

- In a contention-based channel access, there is no central coordination, limited bandwidth and location-dependent contention.
- Explicitly supports time-sensitive traffic such as voice and video from the MAC protocol.

Resource Reservation

- The QoS are defined by three parameters: bandwidth, delay and jitter.

- It requires the reservation of resource such as bandwidth, buffer space, and processing power.
- Due to the mobility of nodes in the Adhoc network the reservation of resources is a difficult task.

Ability to Measure Resource Availability

- The MAC protocol provides *resource availability* to every node based on call admission control and this can also be used for congestion-control decision.

Capability for Power Control

- The energy consumption at the nodes can be reduced by *transmission power control* and it also decreases interferences and increases the frequency reuse.

Adaptive Rate Control

- It refers to the *variation in the bit rate*.
- If a MAC protocol uses, adaptive rate control when the sender and receiver are nearby, it achieves a high data rate and it adaptively reduces the data rate when they are far away.

Use of Directional Antennas

- Directional antennas have many advantages: Increased spectrum reuse, Reduction in interference and Power consumption.
- But most of the MAC protocol uses omni directional antennas.

Routing

- A routing protocol is used to exchange the routing information and to find the feasible path to destination.
- Hop length, minimum power required, information about the path break, minimum processing power and bandwidth are used for mending the broken path.

Mobility

- The important property of Adhoc network is the *mobility of nodes*.
- Due to the mobility of nodes there may be frequent path break, packet collision, transient loop, stale routing information and difficult resource reservation.

Bandwidth Constraint

- The channel is shared by all nodes in the network; the bandwidth available depends on the number of nodes and the traffic they handle.

Error-prone and Shared Channel

- In a wireless channel the bit error rate (BER) is higher when compared to wired networks.
- This is takes into account to design a routing protocol for a Adhoc network.

Location-Dependent Contention

- The wireless channel load varies when the number of nodes increases.
- The number of nodes increases the contention high in the channel.
- The high contention and collision in the channel leads to wastage of bandwidth.

Other Resource Constraints

- Resource constraints consist of computing power, battery power, and buffer storage and also the capability of the routing protocol.
- The major requirement of routing protocol in a adhoc network is
    - Minimum route acquisition delay
    - Quick route reconfiguration
    - Loop-free routing
    - Distributed routing approach
    - Minimum control overhead
    - Scalability
    - Provisioning of QoS

- Support for time-sensitive traffic
- Security and privacy

Multicasting

- Multicasting is important in many applications of Adhoc networks.
- The nodes that form the group to carry out the tasks that require point-to-multipoint and multipoint-to-multipoint voice and data communication use multicasting.
- Multicasting is very challenging for the mobility of nodes with the constraints of *bandwidth and power source*.
- A *tree-based multicasting structure* frequently needs to readjust the broken link and it is highly unstable.
- Such tree-shaped topology leads to high multicasting efficiency and low packet deliver ratio because of frequent tree breaks.
- The multiple links among the nodes in the network result in *mesh-shaped topology*.
- It works well in a high mobility environment.

Robustness

The multicasting routing protocol is used to recover and reconfigure quickly when the link breaks due to mobility of nodes.

Efficiency

Minimum number of transmissions is required to deliver the packet to all the group members.

Control Overhead

Scarce bandwidth in Adhoc networks demands for minimal control overhead.

Quality of Service

QoS supports the multicast routing because the data transferred in a multicast session is time-sensitive.

Efficient Group Management

Group management should perform with a minimum exchange of control message. Group management refers to the process of accepting the session member and maintaining the connectivity among them until the session expires.

Scalability

The multicast routing protocol should be able to scale for a network with a large number of nodes.

Security

Authentication of session members and prevention of non-members from gaining unauthorized information play a major role in military communication.

Transport Layer Protocols

- A transport layer protocol includes setting up and maintaining *end-to-end connections, reliable end-to-end delivery of data packets, flow control* and *congestion control.*
- A reliable connection-oriented transport layer protocol such as transmission control protocol in an Adhoc wireless network arises due to frequent path breaks, stale routing information, a high channel error rate and frequent network partitions.
- Frequent path break occurs in the network due to *mobility of nodes and limited transmission range to the destination node.*
- Depending upon the routing protocol, each path leads to route reconfiguration.
- To find an *alternate path or reconfiguration of broken path* might be longer than the retransmission of packet from the sender.
- The retransmission of packet executes the *congestion control algorithm.*
- This algorithm will decrease the size of congestion window resulting in low throughput.
- The reconfiguration of broken path and the use of the route cache will result in stale routing information.
- The packets are forwarded to the destination node through multiple paths, causing an *out-of-order packet.*
- It forces the receiver to generate the *duplicate ACK* to sender.
- On receiving the duplicate ACK the sender involves the congestion control algorithm.

- The channels are unreliable due to high probability of error caused by interference.
- When the TCP ACK is delayed for more than the round trip time then the congestion control is involved.

Pricing Scheme

- Adhoc network function depends on the *relay nodes*.
- If a node density is sufficient enough to transmit the packet to the network, the relaying neighbor node is not interested in relaying or call i.e., the power is down in that node.
- Assumes the optimal route from node A to node B that passes through node C.
- If node C is not powered on, then node A should have to set up the non-optimal path to node B.
- The *non-optimal path* consumes more resource and affects the throughput of the system.

Quality of Service

- QoS is a *performance level service* offered by a service provider or the network to the user.
- QoS often requires negotiation between the host and the network, which is obtained through resource reservation schemes, priority scheduling and admission control.
- QoS in adhoc network can be based on per flow, per link or per node basis.

QoS Parameters

- The level of QoS and QoS parameter is different from application to application.
- For example in multimedia application, the QoS parameters are bandwidth and delay, whereas military applications need the security and reliability parameter.
- In case of defense applications the trustworthy intermediate host and route through them can be a QoS parameter.

- For hybrid wireless application the availability of link, channel utilization, delay and bandwidth are the key parameters for QoS.
- Communication among nodes in sensor networks requires minimum energy consumption because the battery life and energy conservation is an important QoS parameter.

QoS-Aware Routing

- QoS aware routing protocol uses QoS parameters to find the path.
- For *routing decision,* the parameters are network throughput, packet delivery ratio, reliability, delay, delay jitter, packet loss ratio, bit error rate and path loss.
- For example, if bandwidth is a QoS parameter, the routing protocol uses the path with necessary bandwidth.
- It also reserves the bandwidth for that particular connection.

QoS Framework

- QoS framework is a *QoS service model* which defines the way in which the user requirements are served by per-session or per-class basis.
- The other key component of QoS framework is to find the *feasible paths* in the network that satisfy user requirements.

Self-Organization

- The important property of Adhoc network is to *organize and maintain the network* by itself.
- The activities that a network needs to perform are self-organization, neighbor discovery, topology organization and topology reorganization.
- In the *neighbor discovery phase*, every node gathers information about the neighbor and maintains the data in the data structure.
- The information about the neighbor is gathered periodically by transmitting *beacon packet.*
- In the *topology reorganization phase*, the topology changes occurs due to mobility of nodes, failure of nodes or complete depletion of power sources of the nodes.

- The topology information is updated by the topology changes.
- The two major activities of reorganization is are.
  - Periodic or aperiodic exchange of topological information
  - Adaptability

Security

- In Adhoc networks, the security of communication is very important for military applications.
- The lack of central coordination in a shared wireless medium leads to more *vulnerable attacks* than in wired networks.
- The attacks in a Adhoc network is classified into
  - Passive attack
  - Active attack
- *Passive attack* refers to the attempt made by nodes to obtain information in the network without disrupting the operation.
- *Active attacks* disrupt the operation in the network.
- The active attack executed by the nodes outside the network is called *external attack* and those performed inside the network are called *internal attack or* compromised nodes.

Denial of Service (DoS)

- The network resource unavailability for other nodes either by consuming the bandwidth or overloading the system is known as DoS.
- A simple scenario for DoS attack is by keeping the target node busy by sending the unnecessary packet.

Resource Consumption

- The source availability of resources in the network makes it easy for internal attack, i.e., aiming to consume the available resource.
- The types of resource-consumption attacks are.
  - Energy depletion
  - Buffer overflow

- **Energy depletion**
  - The nodes in the network are highly energy constrained.
  - This type of attack depletes the battery power of nodes by directing *unnecessary traffic* through them.
- **Buffer overflow**
  - This attack is carried out by filling the routing table by unwanted entries or data packet buffer-space with unwanted data.
  - It leads to many problems.

Host Impersonation

- A *node can act as another node* by responding to the control packets to create wrong route entries and terminates the traffic intended to destination node.

Information Disclosure

- A *compromised node acts as informer* by disclosing the confidential information to the unauthorized nodes.

Interference

- A common attack in defense applications is *interference*.
- This is to jam the wireless communication using a single wide-band *jammer*.

Addressing and Service Discovery

- Addressing and service discovery in the network is caused because of *absence of the centralized coordinator*.
- The allocation of non-duplicate addresses to nodes is done by *auto-configuration*.
- In order to maintain the unique address throughout the connected part of the network, the address-detection mechanism is used in case of frequent partitioning and merging of the network.
- Every node in the network should be able to locate the service provided by other nodes.

- Topology changes force to change the location service.
- Identifying the current location of service provider is done by integrating service discovery with route-acquisition mechanism through a viable alternative.

Energy Management

- The lifetime of the network will be enhanced by managing the source and consumption of energy in a node or network.
- *Energy discharge pattern of node* is enhanced by battery life, minimum total energy consumption in finding the route, use of distributed scheduling and handling the processor and interface device to minimize power consumption.
- Energy management is divided into,
  - Transmission power management
  - Battery power management
  - Processor power management
  - Device power management

Transmission Power Management

- In a mobile node the power consumed by radio frequency (RF) is determined by many factors such as state of operation, transmission power, and the technology.
- The transmission, receive and sleep mode of operation is determined by the state of operation.
- Reachability requirement of the network is called transmission power.
- The design of RF hardware should minimize the power consumed by three states.
- When there is no transmission or receiving of packet, it goes to sleep mode and it is wokeup by control signal.

Battery Power Management

- The node battery life is extended by taking advantage of the chemical properties, discharge pattern and selection of battery from set of batteries available for redundancy.

Processor Power Management

- The processor parameter affects the power consumption by executing the number of instructions per unit time and clock speed.
- Based on the load condition, the CPU is turned off and turned on by interrupts after the detection of other events.

Device Power Management

- A mobile node significantly reduces power consumption by *intelligent device management.*
- This is done by the operating system through power down when it is not in use.

Scalability

- The number of nodes in Adhoc network does not grow in the same magnitude as in the internet.
- For mainstream wireless communication the Adhoc networks that include wireless mesh network are installed in widespread.

Deployment Consideration

- The deployment of Adhoc networks is different from that of the wired networks.
- The actual deployment of network is done after a time-consuming plan stage.
- The laying of copper or fiber cable is difficult in reconfiguration or any partial deployment than in Adhoc network.
- The deployment of Adhoc network has following benefits
  - Low cost of deployment
  - Incremental deployment
  - Short deployment time
  - Reconfigurability

Low Cost of Deployment

- Use of multi-hop wireless network eliminates the laying of cable and maintenance as in traditional network.
- Hence the cost is lower than that of the wired network.

Incremental Deployment

- The Adhoc networks are deployed incrementally in geographical regions.
- The deployed networks start functioning immediately after the minimum configuration is done.

Short Deployment Time

- The deployment time is considerably less when compare to wired networks due to the absence of wired links.
- Wiring in urban regions is difficult and time-consuming.

Reconfigurability

- The cost of reconfiguration is high in wired networks when compared to Adhoc networks while covering the same service area.
- The deployment of Adhoc networks varies for different types of application and environment.
- The major issues that are considered while deploying Adhoc networks
    - Scenario of deployment
    - Required longevity of network
    - Area of coverage
    - Service availability
    - Operation integration with other infrastructure
    - Choice of protocol

CONCLUSION

- *The information transferred between two or more points that are not connected through links is called wireless communication.*

- *Radio signals are affected by the objects on the path or media through which they travel. The path by which the radio signal propagates controls the level and quality of signal it received.*
- *When a radio wave falls on any other medium having different electrical properties, a part of it is transmitted into it, while the rest are reflected back.*
- *When the ray hits the edge of an object, the ray bends at the edges of the object there by propagating in different directions. This causes diffraction.*
- *Scattering occurs when the ray travels through a medium, which contains many objects whose dimension is small when compared to the wavelength.*
- *The characteristics of a wireless channel are path loss, fading, interference and Doppler Shift.*
- *The ratio of the power of the transmitted signal to the power of the same signal received by the receiver on the given path is said to be path loss.*
- *The fluctuations in signal strength when received at the receiver is referred as fading.*
- *Fast fading refers to the rapid fluctuations in the amplitude, phase or multipath delays of the received signal due to interference between multiple versions of the same transmitted signal arriving at the receiver at different times.*
- *When objects partially absorb the transmission between the transmitter and receiver it is called slow fading.*
- *In a adjacent channel interference, the signals in nearby frequencies have component outside their allocated range and these components may interfere with on-going transmission in the adjacent frequencies*
- *Co-Channel interference is due to other nearby systems using the same transmission frequency.*
- *Inter-symbol interference refers to distortion in the received signal caused by the temporal spreading and consequent overlapping of individual pulse in the signal.*
- *Doppler shift is defined as the change/shift in the frequency of the received signal when the transmitter and receiver are mobile with respect to each other.*
- *Frequent path break occurs in the network due to mobility of nodes and limited transmission range to the destination node.*

- *Passive attack refers to the attempt made by nodes to obtain information in the network without disrupting the operation.*
- *Active attacks disrupt the operation in the network.*
- *Cell sectoring, cell resizing, and multi-tier cell are techniques that have been used to increase the capacity of the cellular network.*
- *Military sensor networks are used to detect and gain information about enemy movement, explosions, etc.*
- *A sensor network has a set of resource constraints such as limited battery power and communication bandwidth.*
- *The deployment consideration of Adhoc networks are installation, operation and maintenance.*

Chapter II

MAC Protocols for ADHOC Networks

Objective

- *To learn about the fundamentals of MAC Protocol*
- *To know the classifications of MAC protocol*
- *To be familiar with the concepts of Contention based protocols with Reservation Mechanisms*
- *To understand the application of Contention based protocols with Reservation Mechanisms*
- *To interpret the concepts of Multi channel MAC-IEEE 802.11*

2.0. MAC Protocol

Wireless channel is a shared medium. MAC coordinates transmission between users sharing the spectrum. MAC protocol prevent collisions while maximizing throughput and minimizing delay.

2.1. Issues in Designing a MAC Protocol

While designing a MAC protocol for Adhoc wireless networks, the following main issues are needed to be addressed,

- Bandwidth efficiency
- Quality of Service support
- Synchronization
- Hidden and Exposed terminal problem
- Error-prone shared broadcast channel
- Distributed nature/Lack of central coordination
- Mobility of nodes

2.1.1. Bandwidth Efficiency

- **Bandwidth efficiency** is defined as the ratio of the bandwidth used for actual data transmission to the total available bandwidth.
- Since the bandwidth is limited in the radio spectrum for communication, the MAC protocol must be designed in such a way as to utilize the bandwidth in an efficient manner.
- The **control overhead** should be as minimized as possible the bandwidth is reduced.
- The MAC protocol must try to maximize bandwidth efficiency.

2.1.2. Quality of Service Support

- In an Adhoc wireless network, where most of the nodes are mobile, it is difficult to provide QoS support.
- Bandwidth is reserved at one point of time and it becomes invalid if the node moves out of the region where the reservation was made.
- QoS support is essential for military communications involving time-critical traffic sessions.
- The MAC protocol for Adhoc wireless network should have some kind **resource reservation** by considering the wireless channel and mobility of the nodes.

2.1.3. Synchronization

- Synchronization is very important for nodes to get the bandwidth reserved.

- To achieve time synchronization among the nodes, the **exchange of control packets** is required.
- The control packet does not consume more network bandwidth.
- The MAC protocol should consider the synchronization between nodes.

2.1.4. Hidden and Exposed Terminal Problem

- The hidden terminal problem refers to the **collision of packets at receiving nodes** because of the simultaneous transmission of packets from nodes that are not in the direct transmission range of the sender, but are in the transmission range of the receiver.
- When both the nodes transmit the packet at same time without knowledge about each other, the collision occurs.
- For example if nodes S1 and S2 transmit the packet to ode R1 at the same time, the packet collides at R1.
- This is because of S1 and S2 are hidden from each other i.e., they are not in the direct transmission range.
- The exposed terminal problem refers to the inability of the node: **the node is blocked because of transmission by a nearby node.**
- For example if node S1 transmits the packet to node R1 the transmission is already in progress, and node S3 cannot transmit the packet to node R2, which is the neighbor of S1 is already in the transmission mode and hence does not interfere in the on-going transmission.
- The hidden and exposed terminal problems are unique in the wireless network.
- This problem reduces **the throughput of the network** when the traffic load is high.
- The MAC protocol should be free from hidden and exposed terminal problems.

2.1.5. Error-Prone Shared Broadcast Channel

- The important factor of MAC protocol is **the broadcast transmission in the radio channel**, i.e., in the direct transmission range all the nodes receive the packet from the sender.

- The node should not transmit when it is receiving data.
- A node should use the shared medium while the medium is not in use i.e., it should not affect the on-going session.
- The possibility of packet collision is high in wireless network because of multiple nodes using the channel simultaneously.
- A MAC protocol should grant channel access to nodes that minimize collisions and also minimize the bandwidth.

2.1.6. Distributed Nature/Lack of Central Coordination

- The Adhoc wireless network has **no centralized coordinators**.
- In cellular networks the base station acts as a central coordinating node and it will allocate the bandwidth to all other mobile terminals.
- But in the Adhoc network, all nodes are moving continuously and there is no central coordinating node.
- By exchanging the control information the node should schedule the channel in the distributed fashion.
- The MAC protocol should ensure that control information exchange causes the overhead in the network.

2.1.7. Mobility of Nodes

- The important factor that affects the performance of the protocol is the mobility of nodes.
- In Adhoc network the nodes are mobile most of time.
- If the node mobility is very high then the bandwidth reservation and exchange of control information is difficult.
- While designing a MAC protocol the mobility factor should be taken into account so that the performance of the system is not affected by the mobility of nodes.

2.2. Classification of MAC Protocols

- In Adhoc wireless network the MAC protocols are classified into several categories based on criteria such as **initiation approach, time synchronization and reservation approach**.
- MAC protocols for Adhoc network are classified into three types,
    - Contention-based protocols
    - Contention-based protocols with reservation mechanism
    - Contention-based protocols with scheduling mechanism

2.2.1. Contention-Based Protocols

- The contention-based protocols follow the **channel access policy**.
- In this protocol, a node does not reserve any resources in prior.
- Whenever it receives a packet from the sender, to access the shared channel, it contends with its neighbor.
- This protocol does not provide QoS guarantees since the nodes are not regular in accessing the channel.
- Random access protocol can be divided into,
    - Sender-initiated protocols - packet transmission initiated by sender
    - Receiver-initiated protocols - packet transmission initiated by receiver
- Sender-initiated protocols are further divided into,
    - Single-channel sender-initiated protocols
    - Multi-channel sender-initiated protocols

Single-Channel Sender-Initiated Protocols

- In this protocol, **the total available bandwidth is used without dividing it**.
- When a node wins access to the channel then it makes use of the entire bandwidth.

Multi-Channel Sender-Initiated Protocols

- In this protocol, **the entire bandwidth is divided into multiple channels**, so that several nodes transmit data simultaneously using the separate channels.
- Some of the protocols use the frequency channel for transmitting the control information.

2.2.2. Contention-Based Protocols with Reservation Mechanism

- Adhoc wireless network needs to support real-time traffic, which requires QoS guarantees to be provided.
- The contention-based protocol does not guarantee the QoS, so this protocol does not support real time traffic.
- In order to support real time traffic, the protocol reserves the bandwidth in prior.
- This protocol is classified into two types,
    - **Synchronous protocols**
    - **Asynchronous protocols**

Synchronous Protocols

- This protocol needs time synchronization among all the nodes in the network, so reservation of a node should be known to all neighbor nodes in the network.
- Achieving of global synchronization is difficult.

Asynchronous Protocols

- Asynchronous protocols use the relative time information for effective reservation.
- This protocol does not require any global synchronization in the network.

2.2.3. Contention-Based Protocols with Scheduling Mechanism

- This protocol focuses on scheduling the nodes to access the channel and packet scheduling at nodes.
- Scheduling based schemes are also used as priorities among the packets that are queued at the nodes.
- The scheduling schemes are battery characteristics, such as remaining battery power considered, while scheduling the nodes for accessing the channel.

2.3. CONTENTION-BASED PROTOCOLS

- The protocol doesn't have any bandwidth reservation mechanism.
- The nodes are ready to contend for the channel, the winning node gains access to the channel.
- The protocol does not guarantee bandwidth and so it cannot be used for real-time traffic.

MACAW: A Medium Access Protocol for Wireless LANs

- This protocol is based on *multiple access collision avoidance* (MACA).
- MACA protocol is proposed to overcome the shortcoming of CSMA protocols, and is used in wireless network.

2.3.1. MACA Protocol

- This protocol is alternative for CSMA protocol used in wired network.
- In CSMA protocol, for the carrier signal the sender could sense the channel.
- If carrier is present, after random period of time it retries.
- In adhoc wireless network, the transmitter and receiver are not nearby each other all the time.
- In such situations, collision is occurs at the receiver node due to simultaneous transmission by hidden terminals.

- In CSMA protocol the bandwidth utilization is less due to exposed terminal problem.
- MACA protocol does not make use of carrier-sensing for channel access.
- It uses the additional signaling packet for channel access: **Request-To-Send (RTS)** and **Clear-To-Send (CTS)** packet.
- If a node wants to send a data packet, first it will send the RTS packet to the receiver node.
- On receiving the RTS packet, if receiver node is ready to receive the data packet, it sends CTS packet to the sender.
- If the sender receives the CTS packet without any error then it will transmit the data packet to the receiver. This mechanism is depicted in the above figure.
- The node uses the **binary exponential back-off algorithm** when transmitted packet is lost, and waits for random interval of time before retrying the algorithm.
- When the collision is detected by binary exponential back-off algorithm, then the node maximizes the back-off window by double.
- The neighbor node of sender hears the RTS packet and it will not transmit for a long time period, so the sender can send the CTS packet.
- The nearby receiver node hears the CTS packet and it will also defer the transmission until the data packet is received.
- Thus, *MACA overcomes the hidden terminal problem.*
- The node that received the RTS will wait for a short period of time till the sender could receive the CTS.
- During the waiting period if no CTS is heard by the node it is free to transmit the packet when the waiting period is over.
- Hence the *exposed terminal problem is overcome by MACA.*
- But still MACA has certain problems.

2.3.2. MACAW Protocol

- MACA uses the *binary exponential back-off mechanism* at the time of starves flow.
- Function of this protocol shown in Figure 2.1

- For example, both nodes S1 and S2 keep generating traffic at high volume.
- A node captures the channel (node S1) and starts transmitting the packet.
- If another node S2 transmits the packet it gets collided and node **back-off window is incremented by the binary exponential back-off algorithm**.
- Then the probability of S2 acquiring the channel is decreased and over a period of time it gets completely blocked.
- To overcome the problem, the back-off algorithm is modified.
- In the modified algorithm the packet header has an additional field that carries the current back-off value of the transmitting node.
- On the receiving node the packet copies the value of back-off counter.
- Thus the mechanism allocates the bandwidth in a fair manner.
- Another problem in the BEB (Binary Exponential Back-off) algorithm is to adjust the back-off counter value very rapidly, after successful transmission and when collision is detected by a node.
- After every successful transmission the back-off count is reset to the minimum value.
- MACAW uses the multiplicative increase and decrease back-off mechanism to prevent large variations in the back-off values.
- Another modification related to back-off mechanism has been made in MACAW.
- In this mechanism, every node maintains multiple queues, one for each data stream and the back-off algorithm runs on every queue.
- If a node is ready to transmit packets, it determines how long it has to wait to transmit the RTS packet to the destination node and corresponding to the node's queues.
- Then it selects the packet which has minimal waiting time.
- In MACA the RTS and CTS packet are used, but MACAW uses a new control packet called **acknowledgement (ACK) packet**. The need of additional packet is because of the following reasons:
  - In MACA protocol, the transmission error recovery lies in *transport layer* i.e., TCP has minimum time period 0.5 sec to recover from the transmission error. But in MACAW, the error can be recovered in the *data link layer* (DLL), because the recovery period of the time is modified to make the *recovery process quick.*

- In MACAW, after completion of successful transmission of packet, the ACK packet is transmitted. If the sender does not receive the ACK packet, it reschedules the data transmission and then the back-off counter is incremented. If The ACK packet is lost in transmission, then the sender transmits the RTS packet to the receiver, but the receiver instead of sending the CTS packet, sends the ACK packet and the sender moves to next transmission.
- In MACA, when the source node transmits the packet, the exposed node is free to transmit the packet. For example when the transmission is between nodes S1 and R1, node S2 is free to transmit. There is no use of RTS transmission by node S2, until it receives the CTS packet from the node R2. The collision is not possible in the node S2 with the packet transmitted by node S1. As a result, node S2 unnecessarily builds the back-off counter. To overcome the above problem, the MACAW protocol uses another control packet called **data-sending (DS) packet**. In the source node, before transmitting the data packet the DS packet is transmitted which carries the information about the duration of the data packet transmission, so the exposed node updates the information regarding the data packet transmission? If an exposed node overhears the DS packet, then it understands that the previous RTS-CTS exchange was successful, and it defers the transmission until the DATA-ACK exchange.
- One more control packet is used in MACAW protocol and that control packet is called **Request-for-request-to-send (RRTS) packet**. For example when transmission is going on between nodes S1 and R1, node S2 wants to transmit to node R2. Nodes R1 and R2 are neighbor nodes. It receives the CTS packet from node R1, so node S2 defers it transmission and node S2 keeps trying to contend for the channel by incrementing the back-off counter. The main reason of failure is the lack of synchronization information at source S2. The MACAW protocol overcomes the above problem by using RRTS packet. For example, on behalf of source S2 the receiver node R2 contends for the channel. If node R2 receives the RTS packet previously but the transmission is going on between S1 and R1, then node R2 waits for next

contention period to send RRTS packet. When neighbor node hears the RRTS packet it should wait for next two successive slots. If source node S2 receives RRTS packet from node R2, then the regular RTS transmission is done between R2 and S2.

- The *operation of MACAW* is shown in the figure, source node is S, receiver is R and neighbor nodes are N1 N2. When the source S send RTS packet, it is overheard by the neighbor node N1 and it refrains the transmission till the node S receives the CTS. Similarly the CTS packet transmitted by node R is overheard so that it defers the transmission until the data packet is received by node R. On receiving the CTS packet, node S immediately sends the DS packet. On hearing the DS packet node N1 defers it is transmission until the data packet is exchanged. Finally node R receives the data packet and it sends ACK to the source node.

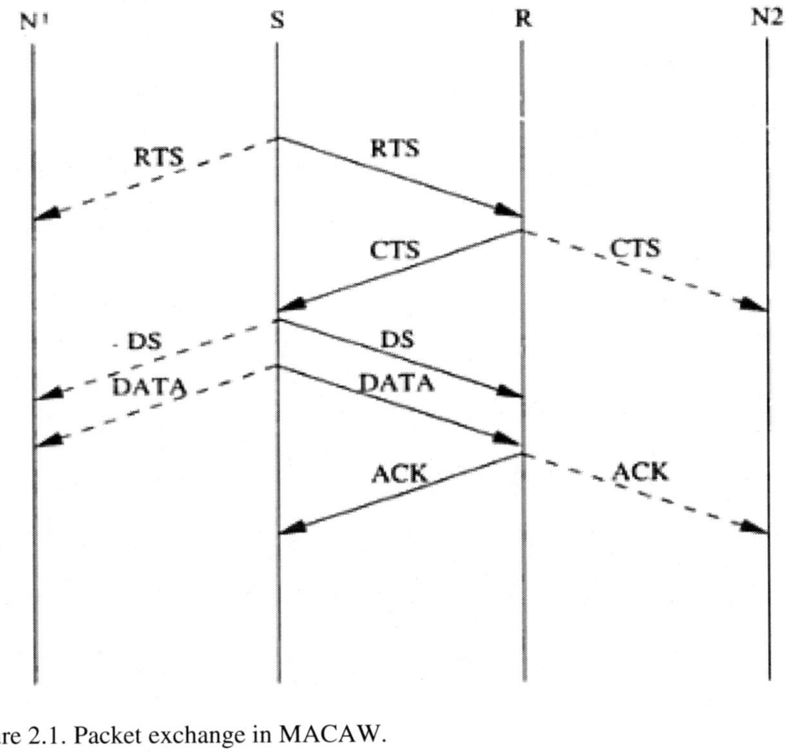

Figure 2.1. Packet exchange in MACAW.

2.4. Contention-Based Protocols with Reservation Mechanism

- The protocols are contention-based, the contention occurs during the resource reservation phase only.
- Once the bandwidth is reserved, the node accesses the reserved bandwidth.
- Hence, QoS support can be provided to real-time traffic.

2.4.1. Distributed Packet Reservation Multiple Access Protocol

- Distributed Packet Reservation Multiple Access Protocol (D-PRMA) extends the centralized packet reservation multiple accesses (PRMA).
- PRMA protocol is used in the distributed scheme and that can be used in ad hoc wireless networks.
- This protocol is proposed to support the voice in a wireless LAN with a base station (BS), where BS is fixed for MAC operation.
- D-PRMA extends to support voice in ad hoc wireless networks.
- D-PRMA is based on the *TDMA scheme*. Here, the channel is divided into fixed and equal-size frames. Each frame has *S slots* and each slot is composed of m *minislots*. The minislot is further divided into two control fields, *RTS/BI* and *CTS/BI* (BI means busy indication). Slots are reserved using the control field and it is used to overcome the hidden terminal problem. For transmitting the packets the first minislot of each slot is contended by all nodes. The remaining (m-1) minislots are given to the node that wins the contention. Until the packet transmission is complete the subsequent minislots are reserved by the winning terminal. If no node wins the first minislot, then the remaining minislots are continuously used for contention, until the contending node wins the minislot. Within the reserved slot, the source and receiver nodes communicate by means of either *time division duplexing* (TDD) or *frequency division duplexing* (FDD). If any node wants to transmit a packet, first it has to reserve the slot, if the slot is not reserved already. It can be reserved by carrier-sensing at beginning of each minislot at a certain period. If a sender node detects the channel to be idle at the beginning of a slot, it sends the RTS packet to the destination through the RTS/BI. On receiving the RTS

packet successfully, the receiver node transmits the CTS packet in respond through CTS/BI of the same minislot. If the sender node successfully receives the CTS packet, current slots are reserved and it can use remaining minislots i.e., minislots from 2 to m. Otherwise, the contention process continues for the subsequent minislots of the same slot.

- The nodes transmitting voice traffic are prioritized over nodes transmitting data traffic and two rules are followed by D-PRMA,
  - The voice nodes are allowed to start contending from minislot1 with probability $p = 1$, data nodes start contending only with the probability $p < 1$. The remaining (m-1) minislots contend with probability $p < 1$ for both voice and data nodes. This is because only after the arrival of voice traffic at the node, the voice node is triggered for the reservation process, this avoids the unnecessary reservation of slots.
  - Only if the node winning the minislot contention is voice node it is permitted to reserve the same slot in each subsequent frame until the end of the session. If a data node wins the contention, then it is allowed to use only one slot, i.e., the current slot and make a fresh reservation for each subsequent slots.
- The nodes that are in radio coverage of the receiver should not permit to transmit simultaneously when the receiver is receiving the packet.
- If it is permitted then the packets collide with the packets of on-going traffic at the receiver end.
- The nodes that are outside coverage area of the receiver are permitted to transmit the packet simultaneously at the same time.
- This means the protocol must be free from the hidden and exposed terminal problem.
- In D-PRMA, if a node wins a contention for minislot1, then the remaining (m-1) slots of the same slot are prevented from other terminals (requirement 1).
- If a slot is reserved for subsequent frames, other nodes are prevented from contending for those reserved slots (requirement 2).
- Trying to satisfy requirement 1, the RTS-CTS exchange mechanism takes place in the reservation process.
- If a node wins the minislot1 then it starts transmitting immediately from minislot2.
- If any other node wants to transmit on the channel, it is busy from minislot2.

- Only when the channel is idle the RTS packet is sent and the other nodes would not contend for the channel until the on-going session gets completed.
- In order to avoid the hidden terminal problem, the receiver transmits the CTS packet and this is heard by all nodes.
- So they defer transmission for the remaining period of that same slot.
- In order to avoid the exposed terminal, if a node hears the RTS it is not allowed to transmit the CTS.
- If the communication is duplex in nature, then it can send and receive simultaneously but exposed nodes are not allowed to transmit.
- A successful reservation cannot be made for minislot if the RTS and CTS packets collide and then the subsequent (m-1) minislots is used for contention.
- To satisfy requirement 2, the receiver transmits a *busy indication* (BI) on the reserved slot through RTS/BI part of minislot1 of the same slot in each subsequent frame.
- The sender also performs the same function by transmitting BI through the CTS/BI part of minislot1 of the same slot.
- If any node hears the BI signal, then the contention for the current frame is not done further.
- Because the reserved slot in each subsequent frame is free from contention.
- The BI signal eliminates the hidden terminal problem.
- The transmitting of BI signal stops when the reservation completes its data transmissions and does not required reserved slot anymore.
- D-PRMA is more suitable for *voice traffic than data traffic applications.*

Floor Acquisition Multiple Access Protocol

- Floor Acquisition Multiple Access Protocol is based on the *channel access* which consists of carrier sensing and collision avoidance between sender and receiver.
- The Floor acquisition is the process of *gaining the control of the channel.*
- Only one node can access the channel at a point of time and it transmits one or more packets to different destinations without any collision.

- The sender senses the carrier by exchanging the RTS-CTS packet.
- The channel is acquired by exchanging the control packet.
- Though the control packets collide with each other, the node sends the data through the channel without any collision.
- This protocol control packet is used to provide the solution for hidden terminal problem.
- FAMA has two variants,
    - RTS-CTS exchange with no carrier sensing
    - RTS-CTS exchange with non-persistent carrier sensing
- To transmit the packet the first variant uses ALOHA protocol and the second protocol uses CSMA protocol.

Multiple Access Collision Avoidance

- In MACA, a node is ready to transmit the RTS packet.
- The neighbor node, on receiving the RTS, defers its transmission for a time specified in RTS.
- On receiving RTS, the receiver node sends CTS back as the respond and waits for a long period of time to get the data packet.
- The neighbor node of the receiver which hears the CTS packet, defers its transmission.
- MACA does not sense the channel.
- The node defers its transmission on receiving RTS and CTS packet.
- The principle of FAMA implies that the data transmission should be collision free and the duration of RTS packet is twice the propagation delay.
- In MACA, the transmissions of burst of packets are not possible.
- But in FAMA-NTR (*Non-persistent Transmission Request*) the MACA is modified to allow the transmission of packet burst using the waiting period of a node.

FAMA-Non-Persistent Transmit Request

- The variant of FAMA is called as **FAMA-NTR** i.e., it combines non-persistent carrier sensing along with the RTS-CTS control packet exchange mechanism.
- Before the sender sending the packet, the node senses the channel.

- If the channel is busy, then the node backs-off and waits for a time period and retries later.
- If the channel is free, then the node sends the RTS packet, the sender listens to channel for one-round trip time to receive the CTS packet from the receiver.
- If the CTS packet is not received within a time period or if the CTS packet is found to be corrupted, then the node backs-off and retries later.
- After receiving the CTS packet without any error, the sender then sends the data packet burst.
- The limited burst is used to maximize the number of data packets, and after the node releases the channel, other nodes contend for that channel.
- Once the channel is acquired by the sender to send the burst of packet, the receiver node waits for τ seconds for processing the each data packet received.
- τ denotes the *maximum channel propagation time*.
- After transmitting any control packet it's enforced to wait for 2 τ seconds.
- This is done to allow the RTS-CTS exchange without any error.
- If node is transmitting the RTS packet, it should wait for 2 τ seconds after transmitting RTS packet in order to enable the receiver node to send the CTS packet.
- After sending the final data packet, a sender node waits for τ seconds in order to allow the receiver node to receive the packet.

Busy Tone Multiple Access Protocols

Busy Tone Multiple Access

- The busy tone multiple access (BTMA) protocol is the earliest protocol and it has been proposed to overcome the *hidden terminal problem* in a wireless environment.
- Function of this protocol shown in Figure 2.2.
- In this protocol the transmission channel is split into
    - Data channel
    - Control channel
- The *data channel* is used to transmit the data packet and *control channel* is used to transmit control packet.

- When a node is ready to transmit, it senses the *channel for the busy tone*.
- If not, it turns on the busy tone and starts the data transmission.
- If the busy tone is present, then the transmission is rescheduled for random rescheduling delay.
- If any other node senses the incoming data channel and allow transmitting the busy tone in control channel.
- Thus, when a node is transmitting, no other node in two-hop neighborhood should transmit.
- *The probability of collision in BTMA is very low and bandwidth utilization is very poor.*

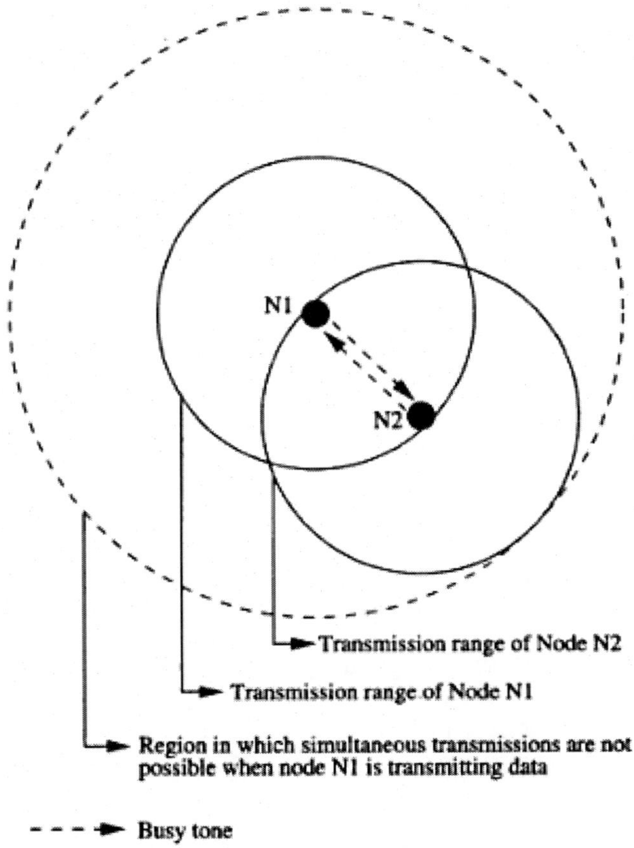

Figure 2.2. Transmissions in BTMA.

Dual Busy Tone Multiple Access Protocol

- The extension of BTMA is a *Dual busy tone multiple access protocol (DBTMA).*
- Again the transmission channel is split into: data and control channels.
- As in BTMA, the *data channel* is used for data packet transmission and the *control channel* is used for control packet and also for transmitting busy tone.
- In DBTMA, the control channel uses two busy tones BT_t and BT_r.
- The BT_t is used by the node to indicate that it is transmitting on data channel.
- The BT_r is used by the node when it is receiving the data packet on data channel.
- The busy tone signals are two sine waves that are different well separated-frequencies.
- When a node is ready to transmit a data packet, it senses the channel to determine if the BT_r signal is active.
- This signal indicates that the neighbor node of ready node or currently receiving the data packets.
- If there is no BT_r signal, then the sender transmits RTS packet on the control channel.
- On receiving the RTS packet, the node that receives the RTS packet checks if BT_t signal is active.
- This signal indicates that the neighbor node is transmitting the packet so it cannot receive the packet at the moment.
- If there is no BT_t signal, then it sends the CTS packet to the sender after receiving RTS packet, the sender node starts sending data packet.
- After completion of the successful transmission, the sender node turns off BT_r signal.
- After receiving all the data packets, the receiver node turns off the BT_t signal.
- When compared with other RTS-CTS based medium access control schemes (MACA and MACAW), *DBTMA exhibits better network utilization.*
- This is because the other schemes block both forward and reverse when transmitting or receiving.

- But DBTMA blocks only reverse or forward when transmitting on the channel.
- *The bandwidth utilization is twice that of RTS-CTS based schemes.*

Receiver-Initiated Busy tone Multiple Access protocol

- The receiver-initiated busy tone multiple access is similar to BTMA, the bandwidth is divided into: *data channel* for transmitting data packets and *control packets* used for transmitting busy tone signal.
- The node transmits the data packet when the busy tone is absent in the control channel.
- The data packet is divided into
    - Preamble
    - Actual data packet.
- The *preamble packet* carries information about the destination node.
- The length of preamble is equal to both data and control channel slots.
- Here the data transmission consists of two steps.
    - The sender transmits the preamble packet.
    - On receiving the preamble the receiver node acknowledges by sending busy tone on the control channel, and then the actual data packet is transmitted.
    - If a sender node wants to transmit the data packet it should wait for a free slot, i.e., the slot in which the busy tone is absent.
    - Once such a slot is found, the sender node transmits the preamble packet on the data channel.
    - On receiving the preamble packet without any error the receiver node sends the busy tone on the control channel.
    - It continues transmitting the busy tone till receiving data from the sender. If the preamble transmission fails, the receiver does not send the busy tone, and then the sender waits for the next free slot and tries again. The busy tone serves two purposes:
        - Acknowledge the sender about the successful reception of the preamble
        - Informs the nearby hidden node about the transmission so that they do not transmit at the same time
- The RI-BTMA protocol is of two types
    - Basic protocol
    - Controlled protocol

- In the basic protocol, nodes *do not have backlog buffers to store data packets*.
- If the packets suffer in collision they cannot be retransmitted.
- Packets cannot be queued at nodes when the network load increases.
- This protocol would work only if network load is low and becomes unstable when the network load increases.
- This problem is overcome in controlled protocol.
- The difference between the two protocols is the availability of backlog buffer at the node in the controlled protocol.
- If a packet suffers in a collision, then the packet is retransmitted and the packets are queued at node.
- If backlog buffer is non-empty then the node is in backlog mode.
- If a node is in backlogged mode, it receives the packet from high layers and puts the packet in buffer and transmits latter.
- Suppose the packet arriving at a node is not in backlogged mode, then the packet is transmitted in current free slot.
- A backlogged node transmits a backlog packet in the next idle slot.
- All packets in backlog buffers wait until transmission succeeds.
- This protocol can work for multi-hop radio network as well as single-hop fully connected network.

MACA-By Invitation

- *MACA-By Invitation* is a **receiver-initiated MAC protocol**.
- MACA-BI reduces the number of control packets compared to the MACA protocol.
- MACA is a sender-initiated protocol which uses the three-way hand shake mechanism, where first RTS and CTS control packets are exchanged followed by actual DATA transmission.
- *MACA-BI eliminates the RTS packet.*
- In MACA-BI, the transmission is initiated by sending ready to receive (RTR) packet to the sender.
- If the sender is ready to transmit, then the actual DATA is transmitted by responds and this protocol uses a *two-way handshake mechanism*.
- The traffic arrival rates of neighboring sender nodes are not known to the receiver node.
- The average arrival rate of packets is estimated.

- For providing the necessary information to the receiver node, the modification is made in DATA packets to carry control information regarding lengths.
- This information is known to the receiver node, the average rate of the flows is easily estimated.
- If the estimation is incorrect or not possible then the MACA-BI protocol is extended by allowing the sender node to send the RTS packet, when RTR packet is not received within a given time period.
- In MACA, to inform the hidden terminal about data transmission, the CTS packet is used, so that they do not transmit at same time.
- In MACA-BI, the role of *CTS is played by RTR packets.*
- The RTR packet carries the information about the time interval when the data is transmitted by its neighbor and the information about the data packet is known to one-hop neighbors or two-hop neighbors i.e., hidden terminal, so they defers transmission during the period.
- Thus, the hidden terminal problem is overcome in MACA-BI and collision of the data is impossible.
- The control packets are still affected by the hidden terminal problem.
- This problem leads to protocol failure, i.e., the RTR packet collides with data packets.
- For example, the RTR packet is transmitted by receiver nodes R1 and R2 collide at node A. So, they are not aware of transmission from S1 and S2.
- When node A transmits the RTR packets, it will collide with data packet at receiver node R1 and R2.
- The efficiency of MACA-BI mainly depends on the ability of the receiver node to predict the traffic arrival rates at the sender nodes.

Media Access with Reduced Handshake

- The Media Access with Reduced Handshake (MARCH) is a receiver-initiated protocol.
- Unlike MACA-BI this protocol does not require any prediction of traffic mechanism.
- This protocol uses the omni-directional antennas to reduce the number of handshakes involved in data transmission.

- In MACA, the RTS-CTS packets exchange for each data transmission, but in MARCH the RTS packet is used only for the first packet of streams and then the CTS packet is used.
- The neighboring nodes overhear the CTS packets transmitted by anode to obtain the information about data packet arrivals.
- Then they send CTS packet to the concerned neighbor node for relaying the data packet from the node.
- In figure above
  (a) The packet exchange in MACA is depicted. It uses two control packets RTS and CTS needed to exchange before data packet transmission. In this figure the node C hears both RTS and CTS packets transmitted by node B.
  (b) MARCH uses the property of broadcast channel to reduce the two-way handshake into single-CTS only handshake. The node B transmits the CTS1 packet, which is heard by node C. Here CTS packet carries the information about the duration of next data packet. Therefore node C determines the time next data packet would be available at node B. It sends CTS2 to receive the data packet from node B. It can be observed that the time taken for a packet transmitted from node A to node D is less in MARCH, t_{MARCH} when compared to MACA, t_{CSMA}.
- In this protocol the CTS packet carries the MAC address of sender and receiver nodes and the route identification (RT_{id}) in that flow.
- The **route identification** is used to avoid misinterpretation of CTS packet and false initiates off CTS-only handshake.
- For example there are two routes, Route 1: A-B-C-D-E and Route 2: X-C-Y.
- When node C hears the CTS packet transmitted by node B, by the RT_{id} field in that packet, the packet is from upstream node on Route1.
- The timer T invokes to set certain time period for node B to receive the packet from node A.
- Once the time expires, node C sends the CTS packet.
- The CTS is overheard by node Y, but its RT_{id} in CTS packet is different from corresponding RT_{id} to Route 2 and node Y does not respond.
- In MARCH, MAC layer maintains the routing table information, but it is not involved in routing.
- The *throughput is significantly high in MARCH when compared to MACA,* while the control overheard is less.

- When the network is heavily loaded, the end-to-end delay in packet delivery is low in MARCH compared to MACA.
- All the above advantages of MARCH are because of low control packet compared to MACA.
- The lower numbers of control a packet reduce the control overhead and less bandwidth is consumed for control traffic.

2.4.2. Collision Avoidance Time Allocation (CATA)

- The Collision Avoidance Time Allocation (CATA) is based on *dynamic topology-dependent transmission scheduling*.
- By *distributed reservation and handshake mechanism* the nodes contend to reserve the time slot.
- CATA supports broadcast, unicast and multicast transmission simultaneously.
- The operation of CATA is based on two basic principles,
  - The receiver(s) of a flow must inform the potential source nodes about the reserved slot on which it is currently receiving packets. Similarly, the source node must inform the destination node(s) about the interference in the slot
  - Usage of *negative acknowledgement* happens for reservation requests and control packet transmission at the beginning of each slot for distributing slot reservation information to senders of broadcast or for multicast sessions
- Time is divided into equal-sized frames and each frame consists of S slots and the slots are further divided into five minislots.
- For transmitting control packets the first four minislots are used and these are called as *control minislots* (CMS1, CMS2, CMS3, CMS4). And for transmitting data packets the last fifth minislot is used and it is called data minislot (**DMS**).
- The data packet is longer than the control packet so the data minislot is much longer than the control minislot.
- During the DMS of the current slot, each node receives the data and it also transmits slot reservation (SR) packet during the CMS1 of the slot.
- This is done so as to inform the other neighbor nodes about the currently active reservation.

- The neighbor node receives the *SR packet* either without error or causes noise at those nodes in both cases preventing such neighbor node from attempting to reserve the current slot.
- The transmission of data in every node is done during DMS of current slot and it also transmits request-to-send (RTS) packet during CMS2 slot.
- If the RTS packet is received by other neighbor nodes or if it collides with other RTS packets at neighbor nodes, it makes the neighbor nodes to understand that the source node is transmitting data during DMS of the current slot.
- Hence the neighbor nodes defer transmission during the current slot.
- The control minislot is used as follows:
    - The intended reservations are made by the sender and it senses the channel to be idle during CMS1 and transmits the RTS packet during CMS2.
    - The receiver node of a unicast session transmits CTS packet during CMS3.
    - On receiving the CTS packet the source node understands the successful reservation and transmits the data during the DMS of the slot and during the same slot in subsequent frames, until the unicast flow gets terminated.
    - Once the reservation of the slot is successful from the next slot onwards, the sender and receiver do not send anything during CMS3 andCMS4 but the sender alone sends not-to-send (NTS) packet.
    - During CMS2 the node receives an RTS packet for broadcast or multicast and if the channel is free during CMS2, it remains idle and does not do anything during CMS3 aqndCMS4.
    - Otherwise, it sends NTS packet during CMS4.
    - The NTS packet serves as a negative acknowledgement, a multicast or broadcast source node receives the NTS packet during CMS4 or it detects the noise during CMS4 and understands that the reservation had failed and does not transmit data during DMS of the current slot.
    - It finds the channel to be free during CMS4 and sends the reservation request and if it is successful it starts transmitting multicast or broadcast packet during DMS of the slot.
    - The length of the frame is very important in CATA.

- The worst-case value of the frame length is the number of slots in the frame which would be Min (d^2+1, N), where d is maximum degree of a node in network, and N is the total number of nodes in the network.
- The CATA works well in single-channel half-dulpex radio.
- It is simple and provides support for collision-free broadcast or multicast transmission.

2.4.3. Hop Reservation Multiple Access Protocol

- Hop reservation multiple access protocol (HRMA) is a *multichannel MAC protocol* based on simple half-duplex, very slow frequency-hopping spread spectrum (FHSS) radios.
- Reservation and handshake mechanism is used to enable a pair of communication nodes to reserve a frequency hop and guaranteeing collision-free data transmission even in the presence of hidden terminals.
- It can also be viewed as a reservation protocol where each time slot is assigned a separate frequency channel.
- From the available L frequency channels, HRMA uses one frequency channel denoted by f_0, used as synchronization channel.
- In f_0 the nodes exchange the synchronization information.
- Remaining L-1 channel is divided into M = $\frac{(L-1)}{2}$ frequency pair denoted by (f_i, f_i*), i = 1, 2, 3…., M.
  - f_i is used for transmitting and receiving the hop reservation (HR) packets
  - Request-to-send (RTS)
  - Clear-to-send (CTS) packets and data packets.
  - f_i*is used for acknowledgement (ACK) of sending and receiving of data packets received or transmitted on frequency f_i.
- In HRMA, time is slotted and the slot is given to separate frequency hop, which is one among the M frequency hops.
- Time slot is divided into four periods: *synchronization period, HR period, RTS period, and CTS period.*
- Each period is used for transmitting or receiving the synchronization packet, FR packet, RTS and CTS packet.

- All idle nodes, that is, currently the nodes that do not transmit or receive packet are hopped together.
- In the synchronizing period of each slot, the idle nodes hop to the synchronizing frequency f_0 and exchange synchronization information.
- They stay idle during the HR, RTS and CTS periods.
- In addition to synchronization period, an **exclusive synchronization slot** is defined at the beginning of each HRMA frame; this slot size is same as other slots.
- The HRMA frame consists of single synchronizing slot followed by M consecutive slots.
- In a network a new node enters and the node remains on the synchronization frequency f_0 for a long period of time to gather synchronization information such as hopping pattern and timing of the system.
- If no synchronization information is received, it assumes that it is the only node in the network, broadcasts the synchronization information and it forms the one-node system.
- The new node entering the system can easily join the network by exchanging synchronization information during every synchronization slot.
- If the length of each slot is μ and μ_s is the length of synchronization period on each slot and the dwell time of f_0 at each frame is $\mu + \mu_s$.
- Consider that where nodes from two different disconnected network partions come nearby, the maximum number of frequency hops M=5 is depicted in the figure.
- The nodes belonging to disconnected network components can easily merge into a single network by overlapping time periods of length μ_s on the synchronization frequency f_0.
- When a node wants to transmit data, it listens to the HR period for the immediately following slot.
- If it hears an HR packet, it backs off for a randomly chosen period.
- If the channel is free during SR period, it transmits an RTS packet to the destination in RTS period and waits for the CTS packet.
- The destination node receives the RTS packet and it transmits CTS packet during CTS period of the same slot and waits for the data packet.

- If the CTS packet correctly reaches the source node, then it implies that the receiver and source nodes are successfully reserved the current slot.
- If source node does not receive any CTS packet, it backs off for random number of slots and repeats the entire process again.
- The source and receiver nodes dwell to the same frequency for the data transmission process, which starts immediately after the CTS period.
- For transmitting acknowledgement a separate frequency is used.
- After the data packet is transmitted, the source node hops to acknowledgement frequency.
- After receiving the data packet, the receiver sends an acknowledgement (ACK) packet back to the source on the acknowledgement frequency.
- The ACK packet transmission/reception is over, the source and receiver nodes go back to reserved frequency to continue the data transmission.
- The transmitted data packet can be of any size.
- Data transmission can be a single packet or train of packets.
- To prevent from hogging onto a particular frequency channel, the maximum dwell period is defined.
- The transmission time of data packet or train of data packets should not exceed maximum dwell time.
- Suppose the source node needs to transmit the data packets across multiple frames, it informs the receiver node through the header of the data packet it transmits.
- On receiving the information the receiver node transmits an HR packet during HR period of the same slot in the next frame.
- The neighbor nodes of the receiver hear the HR packet and refrains from frequency hop reserved.
- On receiving the HR packet, the source node sends the RTS packet during RTS period and other neighbor nodes do not interfere on the reserved frequency hop.
- Both the source and receiver wait for CTS period and transmit the data packet if CTS period gets over.

2.4.4. Soft Reservation Multiple Access with Priority Assignment

- Soft reservation multiple access with priority assignment (SRMA/PA) was developed to support the *integrated services of real-time and non-real-time application* in adhoc wireless networks, maximizing the statistical multiplexing gain.
- In order to contend for and effect reservation of time slots, nodes use collision-avoidance mechanism and soft reservation mechanism.
- On an *on-demand basis* the radio resources are taken over from another node in a non-real-time application, the soft reservation mechanism allows any urgent nodes to transmit the packets that are generated by real-time application.
- SRMA/PA is based on *TDMA protocol,* in this protocol the nodes are allocated different time slots so the transmission is collision-free.
- The main features of SRMA/PA are a unique frame structure and soft reservation capability for distributed and dynamic slot scheduling and time-constrained back-off mechanism.
- Time is divided into frames, each frame consisting of fixed number (N) of time slots.
- Each slot is further divided into six different fields:
  - SYNC
  - Soft Reservation (SR)
  - Reservation request (RR)
  - Reservation confirmation (RC)
  - Data sending (DS) and
  - Acknowledgement (ACK).
- The process of soft reservation is followed by the following
  - For synchronization purpose the SYNC field is used.
  - For transmitting and receiving the corresponding control packets the SR, RR, RC and ACK are used.
  - The DS field is used for data transmission.
  - The SR packet is used as a busy tone and used also to inform the neighbor node about the reservation slot of the transmitting node.
  - The SR packet carries the priority value assigned to the node that has reserved the slot.
  - If an idle node receives the data packet for transmission, the node must wait for a free slot and transmit the RR packet and the node determines whether the slot is free through SR field of the slot.

- If in case of voice terminal nodes, the node tries to take control of the slot reserved already by a data terminal and finds the priority level to be higher than that of data terminal.
- Due to this process the SRMA/PA is different from other protocols where if a node has low priority level compared to other ready nodes, it proceeds to complete the data transmission once it has reserved the channel.

- Based on *service classes and in a static manner*, the priority levels are initially assigned to nodes.
- Once the channel is acquired by a node, the corresponding slot stays reserved until the node completes the transmission.
- For voice and data terminals the node is assigned a pre-specified priority, $p_v^{(R)}$ or $p_d^{(R)}$ respectively, R denotes that the node is already a reserved node.
- To the delay sensitive voice applications it is required that $p_v^{(R)} > p_d^{(R)}$.
- Whenever the reservation fails due to collision, the access priority of the node is updated based on the urgency of the packets.
- *Active state* of a node is said to be when the node is currently transmitting.
- A node is said to be in idle state if it does not have any packet to transmit.
- The active state of a node can be of two states: *access state and reserved state*.
- **Access state** is the one in which the node is backlogged and is trying to reserve a slot for transmission.
- If the slot for transmission is already reserved, then the node is said to be in reserved state.
- Whenever the priority level of voice terminal is in access state, it becomes greater than that of the data terminal in reserved state and this is known from SR field and the corresponding slot is taken over by prioritized voice terminal.
- In this mechanism the value of priority levels such that, $p_v^{(R)} > p_v^{(n)} > p_d^{(R)} > p_d^{(n)}$, where $p_v^{(n)}$ and $p_d^{(n)}$ are the access priority values of a voice and data terminal after its nth reservation attempt results in collision.
- This soft reservation feature of SRMA/PA indicate that whenever the data terminal is reserved where a voice terminal has taken over the

slots because of the nature of its traffic, and the access priority becomes higher than that of the data terminal.

- The RR-RC-DS-ACK exchange mechanism in SRMA/PA is similar to the RTS-CTS-DATA-ACK mechanism of MACAW.
- In SRMA/PA mechanism, the RR and RC packets eliminate the hidden terminal problem.
- The major difference between **SRMA/PA and CATA** is that, for slot reservation in CATA the SR packet is sent by the receiver of the session, whereas in SRMA/PA it is sent by the source node. This feature of SRMA/PA is absent in CATA.
- The access priorities of nodes are updated in a distributed and dynamic manner. This allows sharing the channel in a dynamic manner.
- An idle node becomes active when receiving a new packet for transmission.
- Now, the access state makes the initial access priority value as $p_v^{(0)}$ and $p_d^{(0)}$ depends on whether it is a voice or data terminal.
- If in the random access attempt, the reservation ends up in collision when transmitting the RR packet, then the access priority of the node is increased as follows:

$$p_v(n+1) = p_v(n) + \Delta p_v,\ p_v(0)= p_v^{(0)}$$

$p_d(n+1) = p_d(n) + \Delta p_d$, $p_d(0)= p_d^{(0)}$, where
Δp_v and Δp_d - incremental access priorities for voice and data services respectively.

- The urgency of the traffic queued at the two types of terminals is given below

$$\Delta p_v = \Delta p_v^{(R)} \times \tau_s/\tau_r,$$

$\Delta p_d = \alpha \times l_Q$, where
τ_s - slot duration
τ_r - residual lifetime for the voice service
l_Q - queue length
α - scaling coefficient.

- In the following constraint the access priority of a voice terminal is higher than that of a data terminal, $p_d^{(0)} < p_d(n) < p_d^{(R)} = p_d^{(max)} < p_v^{(0)} < p_v(n) < p_v^{(R)} = p_v^{(max)}$
- In SRMA/PA, the dynamic assignment and update of access priority values are followed by, while the nodes same priority and carrying traffic of the same types will end in collision and *the collision cannot be avoided completely*.
- The collision occurs during the RR field of the slot.
- The collision is avoided by the *binary exponential back-off algorithm* and is used for non-real-time connections.
- A modified binary exponential back-off algorithm is used for real time connections.
- The back-off window is divided into two different regions, where each region has a length of N_{B1} and N_{B2} for real-time and non-real-time traffic.
- Before head-of-line packet each node checks for laxity.
- If the laxity exceeds the threshold T_{limits} slots, then the one slot out of N_{B1} is selected randomly. Otherwise one slot out of N_{B2} is chosen randomly.
- The RR packet is transmitted on the chosen slot.
- If more than one node selects the same random slot and the RR packet collides again, a new back-off window starts immediately after the current slot.

2.4.5. Five –Phase Reservation Protocol

- Five-Phase Reservation Protocol (FPRP) is a *single-channel time division multiple access based broadcast scheduling protocol.*
- To acquire the time slots the node uses the *contention mechanism*.
- This protocol is fully distributed i.e., the multiple reservations can be made simultaneously throughout the network.
- Nodes need not wait for time slot reservation and no ordering among the nodes is followed.
- The *Five-phase reservation process is used for slot reservation*.
- The reservation process is localized; the nodes located within two-hop radius are involved in the process.

- This protocol is *scalable* because the protocol is insensitive to network size and ensures that no collisions occur due to the hidden terminal problem.
- Time is divided into frames.
- The two types of frames are
  - Reservation frame (RF)
  - Information frame (IF)
- In this protocol each RF is followed by a sequence of IFs.
- Each RF has N reservation slots (RS) and each IF has N information slots (IS).
- To reserve an IS, the node needs to contend for the corresponding RS.
- Based on the contentions, in RF a TDMA schedule is generated and it is used in subsequent IFs until the next RF.
- Each RS is made up of M reservation cycle (RC).
- Within each RC, *a five-phase dialog* takes place for nodes to reserve slots.
- If a node wins the contention in RC it reserves the IS corresponding to current RS in the subsequent IF of the current frame.
- Otherwise the node contends the RCs until itself or any other node succeeds.
- In IS the node would be in one of the following states:
  - Transmit (T)
  - Receive (R)
  - Block (B)
- This protocol is free the from hidden-terminal problem, and once the reservation is made with high probability, the node gets sole access to the slot within its neighborhood.
- The protocol assumes that global time is available to all nodes.
- Each node therefore knows when the five-phase cycle would start.
- The reservation process of five-phases are as follows,
  1. **Reservation request phase:** The nodes that need to transmit the packet send reservation request (RR) packets to their destination nodes.
  2. **Collision report phase:** During the reservation request phase if any node detects the collision, then that node broadcasts the collision report (CR) packet. The corresponding source nodes, upon receiving the CR packet take necessary action.

3. **Reservation confirmation phase:** If a source node wins the contention for the slot and does not receive any CR message in previous phase, to confirm the reservation request made in the reservation request phase, the reservation confirmation (RC) message is sent to the destination node in this phase.
4. **Reservation acknowledgement phase:** In this phase, the destination node acknowledges the source node. On the reception of RC it sends back the reservation acknowledgement (RA) message to the source node. The node receives this message and defers its transmission during the span of the reserved slots.
5. **Packing and elimination (P/E) phase:** In this phase two types of packets are transmitted: *packing packet* and *elimination packet.*

The above five-phases are described below.

Reservation Request Phase

- In this phase, each node that needs to transmit packets send an RR packet to the intended destination node with contention probability P, to reserve an IS.
- The node that send RR packet is called requesting node (RN).
- Other nodes keep listening during this phase.

Collision Report Phase

- If any node detects collision of RR packets transmitted in the previous phase, it broadcasts the *collision report* (CR) packet.
- On listening to the CR packet in this phase, an RN comes to know about collision of the RR packet it had sent.
- If the RN does not hear CR packet in this phase, then it assumes that the RR packet did not collide in its neighborhood and then it becomes a *transmitting node* (TN).
- If the node becomes a transmitting node then it proceeds to the next phase, the reservation confirmation phase.
- On the other hand, if the RN hears the CR packet in this phase, it will wait until next RR phase and try again i.e., if two RNs are hidden from each other, then the RR packet collides, both receive the CR

packet and node reservation is made, thereby eliminating the hidden-terminal problem.

Reservation Confirmation Phase

- If the RN node does not receive any CR packet in the previous phase, then it is a TN, sends the RC packet to the destination node.
- The neighbor node receives this packet and understands that the slot has been reserved and it will defer its transmission during corresponding information slots in the subsequent information frames and wait until the next reservation frame.

Reservation Acknowledgement Phase

- On reception of the RC packet, the intended receiver node responds by sending the RA packet back to the TN.
- This is used to inform the TN that the reservation has been established.
- If the TN is isolated and is not connected with any other node in the network then that node would not receive the RA packet and it knows that the node is isolated.
- Thus it prevents isolated nodes from further transmission.
- This phase also serves another purpose.
- Two-hop neighbor nodes that receive the RA packet get blocked for transmitting.
- Then it doesn't disturb the transmission made in the reserved slots.
- Deadlock condition occurs, when more than two TNs are located nearby.
- When there is no common neighbor node present, the RN transmits the RR packets.
- Collision is not reported in the next phase, so each node becomes a TN.
- Deadlocks are of two types:
    - Isolated
    - Non-isolated
- An *isolated deadlock* means where none of the deadlocked nodes is connected to non-deadlocked nodes.

- An *non-isolated deadlock* means at least one deadlocked node is connected to a non-deadlocked neighbor node.
- This phase resolves the isolated deadlock. If none of the nodes transmits the RA, then the TNs abort their transmission.

Packing and Elimination (P/E) Phase

- In this phase, each node sends the packing packet (PP) that is located within two-hops from TN, then it makes the reservation in the previous P/E phase.
- A node receiving the PP understands that there is success in the slot reservation three hops from it and because of some neighbors having blocked for the slots.
- The node takes the advantage and adjusts the contention probability P, so that convergence is faster.
- In order to resolve the non-isolated deadlock, each TN requires transmitting an elimination packet (EP) in this phase, with probability 0.5.
- A deadlock TN, before transmission of its own EP, receives the EP from deadlock TN and knows about the deadlock.
- It backs off by making the slots reserved and does not transmit further in that slot.
- Function is explained in Figure 2.3

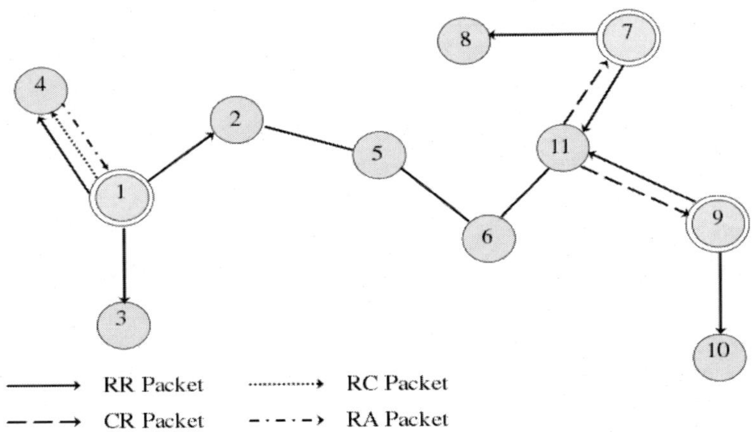

Figure 2.3. FPRP-Example.

- Consider an example, here the nodes 1,7,9 are ready to transmit the packet to nodes 4,8 and 10 respectively.
- During the reservation request phase, all three nodes transmit the RR packets.
- No other node in the two-hop neighborhood of node 1 transmits simultaneously.
- So node 1 does not receive any CR message in the collision report phase.
- Node 1 transmits an RC message in the next phase, for node 4 to send back an RA message and the reservation is established.
- Node 7 and node 9 both transmit the RR packet in the RR phase.
- Here node 9n is within two-hop from node 7.
- So the node 7 and 9 transmit the RR packet simultaneously and it is collide at common neighbor 11.
- Node 11 sends CR packet to both the nodes 7and 9.
- On reception of CR packet nodes 7 and 9 stop contending for the current slot.

2.4.6. MACA with Piggy-Backed Reservation

- MACA with piggy-backed reservation (MACA/PR) is a protocol used for *real-time traffic and to support multi-hop wireless networks.*
- The MAC protocol is based on the MACAW protocol with provisioning of non-persistent CSMA.
- The components of MACA/PR are:
    - MAC protocol
    - Reservation protocol
    - QoS routing protocol.
- This protocol is used to differentiate the *real-time packets from best-effort packets.*
- For real-time packets it provides guaranteed bandwidth and for best-effort packets it provides reliable transmission.
- To define the slots, reservation is made at the nodes.
- Each node maintains the reservation table (RT) and a record of all reserved transmit/receive slot/window of all nodes within the transmission range.

- A MACAW based MAC protocol is used to transmit the non-real time packet.
- The ready node waits for the free slot in RT.
- If it finds a free slot, it again waits for an additional random time for single-hop round-trip delay time, and then senses the channel.
- If the channel is free, then the node transmits an RTS packet to the receiver and if the receiver is ready to receive the packet it responds with CTS packet and then the node sends the DATA packet to the receiver.
- If the receiver receives the DATA packet without any error, it finally it sends the ACK packet to the source.
- The RTS and CTS control packet contains information about the time duration in which the DATA packet is transmitted.
- If any nearby nodes hear these packets, they avoid transmission at that time.
- After the random waiting time, the channel is busy; the node waits for the channel and then repeats the same procedure.
- For the *real-time traffic*, the MACA/PR functions as follows:
    - At certain regular time interval i.e., every CYCLE time period, the sender assumes to transmit the real-time packet.
    - The first data packet of the session is transmitted in the usual manner and a best-effort packet would be transmitted.
    - The source node sends the RTS packet to the receiver.
    - On reception of the RTS packet, the receiver node responds with a CTS packet and then the source node sends the first DATA packet in real-time session.
    - Reservation of the next DATA packet is *piggy-backed* with the current DATA packet.
    - On receiving the DATA packet the receiver node updates the reservation table with the piggy-backed reservation information and then sends the ACK to the source node.
    - The ACK packet is used to confirm the reservation request that was piggy-backed with the DATA packet.
    - The neighbor node hears the DATA and ACK packets update their reservation tables with the reservation information and refrain from transmitting.
- Unlike MACAW, MACA/PR does not use the RTS/CTS packet for transmission of subsequent DATA packets.

- The source node directly sends the DATA packet after receiving the ACK packet and schedules transmission time for the next DATA packet.
- Real-time data transmission occurs as a series of *DATA-ACK packet exchange*.
- The real-time packet is transmitted only once.
- If for a DATA packet the ACK is not received, then the source node just drops the packet.
- The ACK packet is used for serving as a renewing of reservation and resolving packet loss.
- If the source node does not receive any ACK packet for a certain number of consecutive DATA packets then it assumes that the reservation is lost.
- Then the source node restarts the real-time session by exchanging RTS-CTS packet on a different slot on the same link or in case of path break the reservation is made in a different link.
- The source node needs to find the free slot that is free at both nodes to transmit the RTS packet to a receiver node.
- MACA/PR uses periodic exchange of reservation slots to maintain consistent information regarding free slots at all nodes.
- The hidden terminal problem is overcome by automatic exchange of periodic table.
- When a hidden terminal receives a reservation table from a node, it refrains from transmission in the reserved slot of a node.
- At every cycle, the reservation table is refreshed to maintain the slot reservation information.
- For a certain number of consecutive cycles, if the reservation is not refreshed, then it is dropped.
- The transmission of packets in MACA/PR is depicted in the following figure.
- The RTS-CTS is exchanged only for the first packet of the session.
- Each DATA packet carries the information of the next DATA packet.
- The exchange of RTS-CTS is not required for the subsequent DATA packets.
- The reservation tables are updated when the DATA packets are received at the neighbor nodes and none will contend for the channel during a reserved slot.

- The current and future state of the channel of each node is reflected in the *network allocation vector* (NAV).
- Among the best-effort packet and real-time packet transmission, the *highest priority* is given to the real-time packets.
- MACA/PR works as a TDM system for real-time packets with a super frame time of cycle.
- The best-effort packets are transmitted at empty slots of the cycle.
- When a new node joins the network, the node initially remains in the listening mode and receives the reservation tables for each of its neighbor and learns about the network and the reservation made in that network.
- After the initial period, the node shift to normal mode of operation.
- The QoS routing protocol is used with MACA/PR is the destination sequenced distance vector (DSDV) routing protocol.
- In the routing process the bandwidth constraint is introduced.
- Each node periodically broadcasts to the neighbor nodes a pair of preferred path i.e., bandwidth value to each destination.
- The number of preferred path is equal to the maximum number of slots in a cycle.
- After this, a node receives a real-time packet with a certain bandwidth requirement that cannot be satisfied using the current available paths, and the packet is dropped and no ACK packet is sent.
- The sender node would eventually reroute the packet.
- The MACA/PR is an *efficient bandwidth reservation protocol* and it supports real-time traffic sessions.
- The important *advantage of MACA/PR* is that it does not require global synchronization among nodes.
- The *drawback of MACA/PR* is that only if the entire RTS-CTS-DATA-ACK exchange is fit then the free slots are reserved.
- Therefore, many fragmented free slots are not being used so the bandwidth efficiency is reduced in this protocol.

2.4.7. Real-Time Medium Access Control Protocol

- In ad hoc wireless networks, for supporting real-time traffic the bandwidth reservation mechanism is provided by the *real-time medium access control protocol* (RTMAC).

- This protocol consists of two components,
  - MAC layer protocol
  - QoS routing protocol
- The real-time extension of the IEEE 802.11 DCF is a MAC layer protocol.
- The QoS routing protocol is responsible for *end-to-end reservation* and *release of bandwidth resource*.
- The MAC layer protocol is of 2 parts:
  - Medium-access protocol for best effort traffic
  - Reservation protocol for real-time traffic.
- A separate set of control packets from effecting bandwidth reservation for real-time packets are ResvRTS, ResvCTS and ResvACK.
- The RTS, CTS and ACK control packets in transmitting are best-effort packets.
- The real-time packets are given higher priority by the wait time for transmitting a ResvRTS packet that is reduced to *half of DCF inter-frame space* (DIFS) and remaining wait time is used for best-effort packets.
- Time is divided into *superframes*.
- From the given figure it is seen that the superframe of each node is not strictly aligned with other nodes.
- By reserving variable-length time slots on superframe the bandwidth reservations are made in nodes and it is sufficient enough to carry the traffic generated by the node.
- The concept of RTMAC is aimed at slot placement in the superframe and is flexible.
- The superframe consists of a number of *reservation slots* (resv-slots).
- The time duration of each resv-slot is *twice that of maximum propogation delay*.
- A block of resv-slots are required for data transmission.
- A set of resv-slots are reserved for transmission of real-time packets.
- A connection slot is said to be the set of resv-slots which are reserved by a node for a connection on a superframe.
- For successive superframes the same connection-slots are used for the transmission of packets..
- Each node maintains the reservation table which contains information such as sender id, receiver id, and starting and ending time of

reservations that are currently active within the direct transmission range.

- In RTMAC, for all reservation purposes the protocol uses the *relative time*.
- When a node receives the relative-time based information, the relative time is converted into *absolute time* by adding its current time maintained in its clock.
- For reservation a *three-way handshake protocol* is used.
- For example, node A wants to reserve a slot in node B, then it sends a ResvRTS packet that contains the relative information of starting and ending of connection-slot to be reserved.
- On reserving those packets from node A, n node B then first checks the reservation table to see whether it can receive on those resv-slots.
- Then node B sends the ResvCTS packet containing the relative time information of the same resv-slot.
- The neighbor node updates the reservation table on receiving the ResvCTS packet.
- Source node A, on receiving ResvCTS packet, responds by sending ResvACK to the node B.
- This packet also carries the relative time information of the reserved slots.
- The *ResvACK* is meant for the neighbor node of the source node which is not aware of reservation of the node to receive the ResvCTS packet.
- Such node updates the reservation table by receiving ResvACK packet and this packet completes the reservation process.
- The bandwidth reservation process is illustrated in the figure, the NAV (network allocation vector) is maintained at each node and it also contains information about current and future state of the channel at a node.
- To reserve the connection-slot, the ResvRTS packet is first transmitted by the sender node.
- On receiving the packet it checks the NAV to find that the requested connection-slot is free.
- And it responds by transmitting ResvCTS packet on the same connection-slot.
- To complete the reservation process, the sender node sends the ResvACK packet to the receiver node.

- The corresponding connection-slot is marked as reserved at both the sender and receiver nodes.
- The real-time session is started once the reservation is made and the packets are transmitted in the reserved connection-slot by means of Real-time DATA - Real-time ACK exchanges.
- The ResvRTS packet is received on the receiver node and if that slot is already reserved by one of it neighbor nodes, then it does not respond with ResvCTS packet to the sender and the receiver node discards the received ResvRTS packet.
- If a node responds with negative or positive ACK packet it causes collision with reservation made by its neighbor.
- The sender node times out and retries later.
- In case the ResvRTS is received on free slot, but the corresponding connection-slot is not free at the receiver node, then the receiver sends the negative CTS to the sender node.
- On receiving the negative CTS packet the sender node reattempts the following same procedure but with another free connection-slot.
- The real-time session gets finished or a *route break* is detected by the sender, then the sender node release the resources reserved for that session by sending a reservation release RTS (ResvRelRTS) packet.
- This packet is a broadcast packet and the nodes on hearing this packet will update their reservation table to free the connection slots.
- If the receiver node receives the ResvRelRTS packet it responds by broadcasting ResvRelCTS packet to free the reservation slots.
- To find an end-to-end path, RTMAC uses the QoS routing protocol and that matches with the QoS requirement of the user.
- An extension of the destination sequenced distance vector (DSDV) is used as a QoS routing protocol.
- The main *advantage* of RTMAC is *bandwidth efficiency*.
- The nodes are operating in the asynchronous mode, so the successive reservation slots may not strictly align with each other.
- Another advantage of RTMAC is the asynchronous mode of operation, where nodes do not require any global time synchronization.

2.5. Contention-Based MAC Protocols with Scheduling Mechanisms

- These protocols focus on *packet scheduling* at nodes and *transmission scheduling* of the nodes.
- Scheduling decisions are based on various factors such as
  - Delay target of packets
  - Laxities of packets
  - Traffic loads at nodes
  - Remaining battery power at nodes.

2.5.1. Distributed Priority Scheduling and Medium Access in Ad hoc Networks

- This protocol has two mechanisms for providing quality of service (QoS) to support the connections in Ad hoc wireless networks.
- The first mechanism is *distributed priority scheduling* (DPS) which piggy-backs the priority node's current and head-of-line packets on the control and data packets.
- From the neighborhood the information is retrieved from such packets and the node builds the scheduling table which determines its rank compared to other nodes in its neighborhood.
- In order to provide an approximate scheduling, this rank is incorporated into the back-off calculation mechanism.
- The second mechanism is *multihop coordination*, which extends the DPS scheme to carry out scheduling over multi-hop paths.

Distributed Priority Scheduling

- This scheme is based on IEEE802.11 distributed coordination function.
- DPS uses the same basic RTS-CTS-DATA-ACK packet exchange mechanism.
- The ready node transmits the RTS packet and carries the priority tag/ priority index for the current DATA packet to be transmitted.

- The delay target for DATA packet can be a priority tag.
- The intended receiver node responds with a CTS packet on receiving the RTS packet.
- On the CTS packet the priority tag is piggy-backed along with the source node id and the priority tag that is copied from the RTS packet.
- The RTS or CTS packets are received from the neighbor nodes; piggy-backed priority tag information is retrieved and leads to an entry for the packet to be transmitted, in the scheduling tables (STs).
- All nodes maintain an ST holding the information about packets.
- The ST entries are ordered according to their priority tag values.
- The head-of-line packet information is piggy-backed on the DATA packet when the source node transmits the DATA packet.
- On the receiver side, it copies the information onto the ACK packet and sends the response to the source node.
- The Neighbor node updates the STs on receiving DATA or ACK packets.
- When the node receives the ACK packet then it removes its entry from ST for the corresponding DATA packets.
- The piggy-backing and update mechanism is illustrated in the above figure.
- Node 1 needs to transmit a DATA packet to node 2.
- Then it transmits the RTS packet carrying piggy-backed information about the DATA packet.
- Node 4 is the neighbor node of node1 and node 2 and its shown in ST(a)
- On hearing the RTS packet, node 4 retrieves the piggy-backed priority information and makes its entry in ST as shown in ST (b).
- The receiver node responds with the CTS packet and the actual DATA packet is sent once it receives the CTS packet.
- On hearing the DATA packet, node 4 makes the entry for head-of-line packet of node 1 in its ST.
- The new updated status of the ST at node 4 is shown in ST(c).
- The receiver node sends the ACK packet to node 1.
- When the ACK packet is heard by node 4, it removes its entry from ST.
- ST (d) depicts end of the data transfer session.
- With respect to the neighbor node the rank of the node is given in each of the node scheduling tables.

- The back-off period is determined by the rank information. The back-off distribution is given by,

$$\text{Back-off} = \begin{cases} Uniform\,[0, (2tCWmin) - 1 \; r = 1, n < nmax \\ \alpha * CWmin + Uniform\,[0, \gamma CWmin - 1] \; r > 1, n = 0 \\ Uniform\,[0, (2n\gamma CWmin) - 1 \; r > 1, n \geq 1 \end{cases}$$

CW_{min} - Maximum size of the contention window
r - Rank in the scheduling table of the node's highest priority packet
n - Current number of transmission attempts made by the node
nmax - Maximum number of retransmissions permitted
α, γ - Constant used to control the congestion in the second attempt for the highest ranked nodes.

Multi-Hop Coordination

- By means of the multi-hop coordination mechanism, the *excess delay incurred by a packet at the upstream nodes is compensated for at the downstream nodes.*
- When a node receives a packet, it would have already received the priority index of the packet piggy-backed on the previous RTS packet.
- In case the node is an intermediate node which has to further forward the packet, the node calculates the *new priority index* of the DATA packet in recursive fashion, based on the received value of the priority index.
- If $d_{i,j}^{\ k}$ is the priority index assigned to the kth packet of flow i with size l_i^k at his j^{th} hop, if t_i^k is the time at which the kth packet of flow i arrives at its first hop, then the new priority index assigned to the received packet at intermediate node j is given as,

$$d_{i,j}^{\ k} = t_i^k + \delta_{i,1}^{\ k}, \; j = 1$$
$$d_{i,j-1}^{\ k} + \delta_{i,j}^{\ k}, \; j > 1$$

- The increment of the priority index $\delta_{i,j}^{\ k}$ is non-negative function of i, j, l_i^k and t_i^k.
- Because of this mechanism, if a packet suffers due to excess delay at the upstream nodes, the downstream nodes increase the priority of the packet so that the packet is able to meet its *end-to-end delay target.*

2.5.2. Distributed Wireless Ordering Protocol

- Distributed wireless ordering protocol (DWOP) consists of a *media access scheme along with scheduling mechanism* and it is based on the *distributed priority scheduling scheme.*
- In DWOP, the packet access the medium according to the order and it is specified by ideal reference scheduler like first-in-first-out (FIFO), virtual clock or earlier deadline first.
- In this protocol, the reference scheduler is FIFO and the control packets are used to piggy-back the priority information regarding head-of-line packets of nodes.
- The concept of DWOP is that a node is eligible to contend for the channel only if its queued packet has smaller arrival time compared to the other arrival time in ST.
- In order to keep the actual schedule close to the reference FIFO schedule, two additional table management techniques are used
    - receiver participation
    - stale entry elimination are used.
- This protocol may not suffer due to information asymmetry.
- In most of the networks the nodes are not in the radio range of each other, a transmitting node may not be aware of arrival time of packets which is queued at another node and they are not in direct transmission range.
- The fair sharing bandwidth might be affected due to information asymmetry.
- For example, in figure (a), the sender of flow B is aware of packets to be transmitted by the sender of flow A.
- Whenever higher priority packet is queued at the sender of flow A, the sender of flow B defers its transmission because of high priority packets.
- But the sender of flow A is not aware of the arrival time of packets that are queued at the sender of flow B and it concludes that it has the highest priority packet in its neighborhood.
- Therefore, node1 continuously tries to gain access to the channel.
- This would result in flow B receiving an unfairly high share of available bandwidth.
- In order to overcome the information asymmetry problem, the receiver participation mechanism is used.

- In the *receiver participation mechanism*, the receiver node uses the ST information to find whether the sender is transmitting out of order and whether that violates the FIFO schedule and an out-of-order notification is piggy-backed to the sender by using control packets (CTS/ACK).
- Whenever the receiver node detects the FIFO schedule violation and send the notification about FIFO violation, the actual transmission may not follow the FIFO schedule, rather it approximates to the FIFO schedule.
- On receiving the out-of-order packet from the sender node, the receiver node transmits the notification about the actual rank R to the sender with respect to receiver's local ST.
- The sender node receives the *out-of-order notification and* goes into back-off state after completing the transmission of the current packet.
- The back-off period is given by,

$$T_{back\text{-}off} = R^*(EIFS + DIFS + T_{success} + CW_{min})$$

- $T_{success}$ is the longest possible time required to transmit a packet successfully, including the RTS-CTS-DATA-ACK handshake.
- The back-off node allows the higher priority packets of the receiver to be transmitted first.
- To obtain a perfect FIFO schedule, the receiver may not reply to the out-of-order request (RTS) of the sender.
- This would cause the sender to time out and back-off and thereby avoiding the out-of-order transmission.
- In DWOP, the node access the medium depending on the rank in the receiver's node ST and information maintained at the ST must be consistent to the network scenario.
- The stale entry elimination mechanism makes sure that the STs are free of stale entries.
- The entry in ST is deleted when the ACK packet for the corresponding entry is heard by the node.
- If the ACK collides at the node, the entry in ST will never be removed.
- This may cause large variation from the ideal FIFO schedule.
- Figure (b) shows the example for a *collision scenario*.
- The sender and receiver of flow B have stale entries because of collisions.

- And this collision is caused by packets belonging to flow A and flow C at the sender and the receiver of flow B.
- In this case, there is a stale entry in the ST of a node and the node has a head-of-line packet position remaining fixed, the entries below head-of-line packets keep changing.
- The *stale entry detection method* is used for this observation.
- When a node observes that the rank remains fixed whose priorities are below the priority of its head-of-line packet being transmitted.
- It concludes that it have many stale entries in ST.
- Assuming it to be the stale entry, the node simply deletes the oldest entry from its ST.
- This mechanism is used to eliminate stale entries from the ST of nodes.

2.5.3. Distributed Laxity-Based Priority Scheduling Scheme

- The distributed laxity-based priority scheduling (DLPS) scheme is based on packet scheduling scheme, where scheduling decisions are made by the state of the neighboring nodes and feedback from destination nodes regarding packet losses.
- The packets are reordered based on their *uniform laxity budgets* (ULBs) and the packet delivery ratios of the flow.
- Each node maintains two tables:
    - Scheduling table (ST)
    - Packet delivery ratio table (PDT).
- The ST maintains the information about packets to be transmitted by the node and packet overheard by the node are sorted according to the priority index values.
- This *priority index* represents the priority of a packet.
- The PDT maintains the count of packets transmitted and count of acknowledgement (ACK) packets received from the node.
- This information is used for calculating the current packet delivery ratio.
- Feedback mechanism shown in Figure 2.4.
- By means of a feedback mechanism, a node keeps track of packet delivery ratios.

- The overall functioning of the feedback mechanism is depicted in the figure.
- According to their arrival times, the incoming packets are queued in the node's input queue.
- The senders sort the incoming packets according to the priority values and insert them into the transmission queue.
- For transmission the highest priority packets are selected.
- After transmitting a packet, the node updates the count of the packets transmitted so far in its PDT.
- On receiving the packets, a feedback is initiated by the count of DATA packets received and it is conveyed to the source through ACK packet traversing in the reverse path.
- These two informations, together denote s_i and it is received by the *feedback handler* (FIH).
- The FIH, in parallel also sends the previous state information s_{i-1} to the *priority function module* (PFM).
- At each node the ULB of the packet is available in the ST.
- This information is also sent to PFM, which uses this information to calculate the priority indices of packets in the ST.
- The packet delivery ratio (PDR) of the flow is computed at any time by using the count of DATA packets transmitted (pktsSent) and count information carried by the ACK packet (acksRcvd).
- The packet delivery ratio of a flow is computed as,

$$PDR = \frac{\text{acksRcvd}}{\text{pktsSent}}$$

- Priority index of a packet (PI) is defined as,

$$PI = \frac{\text{PDR}}{\text{M}} \times \text{ULB}$$

$$ULB = \frac{\text{deadline} - \text{curretTime}}{\text{remHops}}$$

M - user-defined parameter for the desired packet delivery ratio of a flow
Deadline - end-to-end deadline target of the packet
Current time - current time of the node's local lock.

- When the number of packets belonging to a flow is greater to meet their delay target, the term $\frac{\text{PDR}}{\text{M}}$ would have a high value. Hence the priority index will be high for the packets of that flow and the actual priority of the packets will be low.
- When a few packets of the flow meet their delay target, the value of $\frac{\text{PDR}}{\text{M}}$ will be much less thereby lowering the priority index and increasing the priority of the packets of that flow.
- ULB also plays an equally important role.
- It carries piggy-backed information regarding the highest priority packet queued at the node by modifying the RTS and CTS packets transmitted by a node.
- If any node hears the packet it retrieves the piggy-backed priority information, calculates the *priority index* and adds its corresponding entry in its *ST*.
- When a node hears the DATA packet, then the entry for the corresponding packet exists in the ST, and the entry is deleted from the ST.
- Only when the ACK for transmitted DATA is received, the sender node deletes its entry from ST.
- In case a DATA packet transmitted may not be heard by a node previously in the transmission range of a sender node with high priority packet.
- This might be because of node mobility and channel errors.
- In this situation the stale entry affects the desired scheduling of packets.
- When the network load is high, some packets miss the deadline targets while waiting in the node's queue.
- This is reason for the stale entries in the ST and such packets will never be transmitted.
- Table updates are performed to remove stale entries; the entries whose deadline targets have been missed already are deleted from the ST.
- The back-off mechanism is used in DLPS to reflect the priority of the node's highest priority packet on the back-off period taken by the node.
- If r is the rank in ST of the node of the current packet to be sent, n is the number of retransmission attempts made for the packet and nmax is the maximum number of retransmission attempts permitted. The back-off interval is given by,

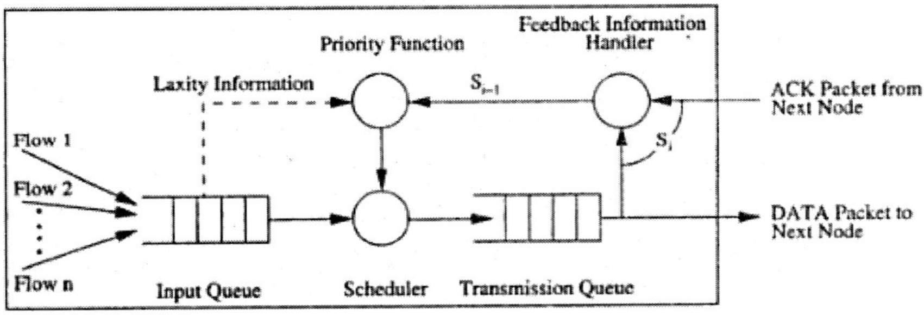

Figure 2.4. Feedback mechanism.

$$back - off = \begin{cases} Uniform[0, (2^n, \text{CWmin}) - 1], \text{if r} = 1 \text{ and n} \leq \text{nmax} \\ \frac{\text{PDR}}{M} \times \text{CWmin} + \text{Uniform}[0, \text{CWmin} - 1], \text{if r} > 1 \; and \; n = 0 \\ ULB \times CWmin + Uniform\ [0, (2^n \times \text{CWmin}) - 1], \text{otherwise} \end{cases}$$

CWmin - Minimum size of the contention window
M - Desired packet delivery ratio.

- This protocol delivers a higher percentage of packets to their delay targets with low average end-to-end delay in packet delivery compared to the 802.11 DCF and DPS scheme.

2.6. MULTICHANNEL MAC PROTOCOL

- In the multichannel MAC protocol (MMAC) multiple channels are used for data transmission and there is no control channel.
- Every node maintains a data structure called *preferable channel list* (PCL).
- The information maintained in PCL is the usage of the channels within the transmission range of the node.
- Channels are classified into three types based on their usage,
    - High preference channel (HIGH)
    - Medium preference channel (MID)
    - Low preference channel (LOW)
- High preference channel (HIGH)

This is the channel selected by the current node and being used by the node in the current beacon interval. Since a node has only one transceiver, there can be only one HIGH channel at a time.

- Medium preference channel (MID)
  Medium preference channel is defined as the channel which is free and it is not being used currently in the transmission range of the node. If there is no HIGH channel available, a MID channel would be the next preference.
- Low preference channel (LOW)
  A channel is currently being used in the transmission range of the node by other neighboring nodes. For each Low state channel a counter is associated with the channel and it also maintains the source-destination pairs for current beacon interval that have been chosen for data transmission.
- In this protocol time id is divided into *beacon intervals* and each node in the network is synchronized by periodic beacon transmissions.
- Every node has beacon interval that starts and ends almost at the same time.
- For every beacon interval, there is a time interval called the *adhoc traffic indication messages* (ATIM) window.
- The nodes in the network use this window to negotiate for channels for transmission. This ATIM messages consist of ATIM, ATIM-ACK (ATIM-acknowledgement), and ATIM-RES (ATIM-reservation) and are used for the negotiation.
- ATIM messages are exchanged in a particular channel called the *default channel*.
- This channel is one of the multiple available channels.
- For sending the DATA packet outside the ATIM window the default channel is used.
- When a node wants to transmit the current beacon interval to the intended destination node the ATIM packet is used and this message also carries the PCL of the transmitting node.
- On receiving the packet the destination node uses its own PCL and PCL is carried on the packet to select the channel.
- This channel information is sent to the source node using ATIM-ACK packet.
- On receiving the ATIM-ACK packet from destination node determines whether the channel is used to transmit the packet as mentioned in the ATIM-ACK message.

- The source node responds to the destination node by sending ATIM-RES packet.
- The ATIM-ACK and ATIM-Res packet are used to inform the neighbor nodes of source and destination nodes that this channel is going to be used for transmission in the current beacon interval.
- The node that hears this packet updates the PCLs.
- At the end of ATIM window, the source and destination nodes start exchanging the RTS/CTS control packets.
- If the source node is not able to use the channel selected by the destination node and if it cannot transmit packets to destination on the current beacon interval, it has to wait for the next beacon interval for negotiating channels.
- Due to collision the ATIM packet is lost and in order to prevent this each node waits for back-off period before transmitting the ATIM packet.
- The operation of MMAC is depicted in the figure below.
- At the beginning of the beacon interval, source node S1 transmits an ATIM message to receiver R1.
- On responding to ATIM message, the receiver node R1 sends the ATIM-ACK (ATIM-ACK (1)) with ID1 of channel it prefers.
- On receiving the node S1 confirms the reservation by sending ATIM-RES packet (ATIM-RES (1)) for channel 1.
- Node R2 overhears the ATIM-ACK (1) packet sent by the receiver R1.
- When node R2 receives an ATIM packet from source S2, with a different channel of ID2, the channel information is sent to source node S2 through the ATIM-ACK (ATIM-ACK (2)) packet.
- Node S2 accepts channel2, responds with ATIM-RES (2) packet and the reservation is established.
- At the end of ATIM window the data transmission (RTS-CTS-DATA-ACK) packets between the node pairs S1-R1 and S2-R2 take place on reserved channel.
- In this protocol, the channel selection is made by the receiver node which plays a dominant role.
- The actual data transmission is protected by the RTS/CTS control packet exchange.
- The node contends for the channel if it is transmitting the packet on the same channel, as in IEEE 802.11 packet transmission.
- The power-saving mode is employed in this protocol.

- The node goes into power-saving doze mode, if the node is neither going to transmit nor going to receive the packets.
- The receiver selects the channel in the following manner. The receiver node uses its own PCL and the received PCL from source node for selecting the best channel for communication with the source node. And this procedure tries to balance the network load on the channels.
- Source node S sends the ATIM packet to receiver node R, and the channel is selected as below,
    - If the HIGH state channel exists in node R's PCL, then that channel is selected.
    - If the HIGH state channel exists in node S's PCL, then this channel is selected.
    - Else if the common MID state channel exists in both of node S and node R's PCL, then that channel is selected. If many such channels exist, one of them is selected randomly.
    - Else if the common MID state channel exists at only one of the two nodes, then that channel is selected. If many such channels exist, one of them is selected randomly.
    - If all channels in both PCLs are LOW state, the counter of the corresponding channels at nodes S and R are added, and the channel with the least count is selected.
- MMAC uses simple hardware and it requires the single transceiver.
- The throughput of MMAC is higher than the IEEE 802.11 when the network load is high.
- Unlike other protocols, the packet size of MMAC is needed to be increased in order to take advantage of the presence of an increased number of channels.

2.7. IEEE 802.11

- IEEE 802.11 is standard for WLANs, which has many vendors of WLAN products.
- The latest version of IEEE 802.11 is 802.11b and this standard is for *Wi-fi* (Wireless fidelity).
- IEEE 802.11 standard deals with physical and MAC layer in WLAN.

- The first WLAN standard is IEEE 802.11 and it faced the challenges of organizing a systematic approach for defining a standard for wireless wideband local access.
- Compared to other LAN standards, wireless standards need to support the mobility of nodes.
- This standard examines the connection management, link reliability management and power management.
- A reasonable duration of time was taken to develop the IEEE 802.11a and IEEE 802.11b enhancements.
- In the IEEE 802.11 standard, the mobile terminals (MTs) are operates in two modes:
  - Infrastructure mode – MTs can communicate with one or more APs connected to WLAN
  - Ad hoc mode – MTs can communicate with each other without using an AP

Physical Layer

- The medium to be used at the physical layer is supported by IEEE 802.11 in three options.
- One is based on *infrared* and other two are based on *radio transmission*.
- The physical layer is conceptually subdivided into two parts
  - Physical layer dependent sublayer (PMD)
  - Physical layer convergence protocol (PLCP).
- PMD is used for encoding, decoding modulation of signals and deals with the particular medium.
- The function of PLCP in the physical layer is to offer a service access point (SAP) that is independent of the transmission bandwidth and a clear channel assessment (CCA) sense signal to the MAC layer.
- The MAC layer uses the CCA to implement the CSMA/CA mechanism.
- The physical layer in the original 802.11 standard follows the three choices:
  - *Frequency hopping spread spectrum* (FHSS) operates in the 2.4 GHz industrial, scientific and medical (ISM) band, at 1 Mbps and 2 Mbps of data rate.

- *Direct sequence spread spectrum (DSSS)* operates in 2.4 GHz of ISM band, at 1 Mbps and 2 Mbps of data rate.
- *Infrared* operates at the wavelength of 850-950 nm range and the data rate is at 1 Mbps and 2 Mbps.

Carrier Sensing Mechanism

- The channel sensing in IEEE 802.3 is very simple when compared to IEEE 802.11.
- In IEEE 802.11 the channel sensing mechanism is performed either physically or virtually.
- The sensing of physical layer is through the *clear channel assessment* (CCA) signal provided by PLCP in the physical layer of IEEE 802.11.
- The CCA is generated through sensing of the air interface or by sensing the detected bits in the air or by checking the received signal strength (RSS) of the carrier against the threshold.
- When the decision is made based on detected bits, it may be more slow and reliable when compared to the decision based on RSS.
- This is because it creates a false alarm by measuring the level of interference.

Basic MAC Layer Mechanism

- The primary function of this MAC layer is to statistically multiplex the transmission request to various wireless stations that operate in an area.
- Wireless transmissions are broadcast in nature and contend to access the shared channel in order to avoid collisions or to reduce the number of collisions.
- The MAC layer also supports *roaming, authentication and taking care of power conservation.*
- The MAC layer supports asynchronous data service and optional real-time service.
- For unicast and multicast packets, the asynchronous data service is used.

- The real-time service supports only infrastructure-based network where APs control access to the shared medium.

Distributed Foundation Wireless Medium Access Control (DFWMAC)

- In IEEE 802.11, distributed coordination function (DCF) is the primary access method and this is the basic function in carrier sense with multiple access and collision avoidance (CSMA/CA).
- The hidden terminal problem is avoided by using the RTS-CTS mechanism.
- To provide real-time service, the second method *point coordination function* (PCF) is used.
- When PCF is in operation, the AP is used to control the medium access and avoid simultaneous transmissions by the nodes.

Inter-Frame Spacing (IFS)

- The time interval between the transmissions of two successive frames by any station is called *inter-frame spacing*.
- IFS is divided into four types:
  - SIFS
  - PIFS
  - DIFS and
  - EIFS from shortest to longest.
- They denote the *priority level of access* to the medium.
- The wait time to access the medium is lower in shorter IFS and it has higher priority to access the medium.

Short Inter-Frame Spacing (SIFS)

- SIFS is the shortest of all IFS and has a higher priority to access the medium.
- This is used for short control messages like acknowledgements for data packet and polling response.

- After the channel is sensed to be idle for a minimum period of atleast SIFS the transmission of any packet then begins.

PCF Inter-Frame Spacing (PIFS)

- The waiting time of PIFS lies between SIFS and DIFS.
- This is used for real-time service.

DCF Inter-Frame Spacing (DIFS)

- The DIFS is used to transmit packets for the station which operates under the DCF mode.
- This is also used for asynchronous data transfer within the contention period.

Extended Inter-Frame Spacing (EIFS)

- It is the longest of all IFSs and has least priority to access the medium.
- This is used for resynchronization whenever the physical layer detects incorrect MAC frame reception.

CSMA/CA Mechanism

- The MAC layer mechanism used by IEEE 802.11 WLANs is the *carrier sense with multiple access and collision avoidance* (CSMA/CA).
- Carrier sense with multiple access and collision detection (CSMA/CD) is also a well-studied technique in IEEE 802.x wired LANs.
- This technique cannot be used in WLANs because the error rate is much higher and collision will lead to drastic reduction in throughput.
- Detecting the collisions in the wireless medium is not possible and for that purpose the collision avoidance technique is adopted.

The Medium Access Mechanism

- If a node wants to transmit the packet, first the medium is sensed to be idle for a duration of DIFS.
- The node backs-off its transmission when the medium is busy, in which case the station defer channel access by a random time is chosen within a contention window (CW).
- The value of CW can vary between CW_{min} and CW_{max}.
- The time intervals are chosen by using propagation delay, delay in the transmission, and other physical dependent parameters.
- If the back-off counter reaches zero and expires, the station can access the medium.
- In the back-off process, if a nodes detects a busy channel it freezes the back-off counter and this process continues when the channel becomes idle for a period of DIFS.
- Each station uses the back-off procedure atleast once in a every successive transmission.

Contention Window Size

- The important parameter of the contention window is the contention window size.
- If the CW is small in size, then there will be high probability of packet collision.
- If the CW is high in size, there will be unnecessary delay because of large back-off values.
- The *binary exponential back-off technique* is used when number of stations that contend for channel access and this is similar to the technique used in IEEE 802.3.
- Each time a collision occurs, the contention window is set to random value between (0, CW_{min}).
- The CW doubles its size up to a maximum of CW_{max}.
- At high load, the CW size is high and the resolution power of the system is also high.
- At low load, the CW size is low and has low access delay.

Acknowledgements

- Acknowledgements (ACKs) are used to ensure the correct delivery of data packsent.
- In unicast packet, the receiver accesses the medium only after waiting for SIFS and sends an ACK while other stations have to wait for DIFS period plus back-off time.
- Because of this the probability of collision is reduced.
- The Higher priority is given for a ACK to be sent for perilously received data packet rather than to start a new data transmission.
- The MAC layer frame uses *cyclic redundancy checksum* (CRC) technique to ensure correct reception.
- Transmission takes place when ACK is not received by the sender.
- The number of retransmissions is limited and if the retransmission count exceeds this limit and the failure is reported to the high layer.

RTS-CTS Mechanism

- In a wireless network, the major problem is the *hidden-terminal problem.*
- The classic example of the problem arising due to incomplete topology information in wireless network.
- In case one node can receive from two other nodes that cannot hear each other. The receiver is bombarded by both the senders, resulting in collision and reduced throughput.
- But the sender is unaware of this and gets the impression that the receiver clearly listens to them without any interference.
- This is called hidden-terminal problem.
- This problem is avoided by using RTS-CTS mechanism as shown in Figure (b) above.

How RTS-CTS Works

- Request to send (RTS) packet is sent by the sent to the receiver.

- The packet includes information about the next data packet to be transmitted to the receiver and the expected time duration of the whole data transmission.
- The packet is received by all stations that can hear the sender.
- After receiving this packet every station sets *network allocation vector* (NAV) accordingly.
- This NAV is used to specify the time when the station is permitted to attempt transmission. After waiting for the SIFS period, the receiver of data packet sends clear to send (CTS) packet to the sender if it is ready to accept the data packet.
- The CTS packet contains duration field and after receiving the CTS packet all stations set NAVs.
- The stations which set NAVs are within the transmission range of the receiver.
- The set of stations which receives the CTS packet may be different from the set of stations that receive the RTS packet, which indicates the presence of hidden-terminal.
- Once the RTS packet has been sent, the CTS packet is successfully received and all the nodes in transmission range of the sender and receiver are informed that the medium is reserved for one sender.
- Then the data transmission is started by the sender after waiting for SIFS.
- On receiving the packet the receiver waits for another SIFS and sends the ACK packet.
- As soon as the transmission is over, the NAV in each node marks the medium as free and the process is repeated again.
- The collision can occur at the beginning of RTS or CTS being sent.
- Once the RTS and CTS packet is transmitted successfully, the node listens to RTS or CTS packet to refrain from causing collision.
- The use of RTS-CTS packet before the data packet transmission is a form of virtual carrier sensing.

Overhead Involved in RTS-CTS

- The medium is reserved prior to a particular data transfer in order to avoid collisions during transfer.
- The transmission of RTS-CTS results in non-negligible overhead.

- The RTS threshold is used to determine whether to start the RTS-CTS mechanism or not.
- If the frame size is more than the RTS threshold, the RTS-CTS mechanism is activated with *four-way handshake* i.e., RTS-CTS-DATA-ACK follows.
- If the frame size is below the RTS threshold, then the nodes resort to a *two-way handshake* i.e., DATA-ACK.

CONCLUSION

- *Bandwidth efficiency is defined as the ratio of the bandwidth used for actual data transmission to the total available bandwidth*
- *Adhoc wireless network should have some kind resource reservation by considering the wireless channel and mobility of the nodes.*
- *The hidden and exposed terminal problems are unique in the wireless network.*
- *The Adhoc wireless network has no centralized coordinators.*
- *In cellular networks the base station acts as a central coordinating node and it will allocate the bandwidth to all other mobile terminals.*
- *The contention-based protocols follow the channel access policy.*
- *PRMA protocol is used in the distributed scheme and that can be used in ad hoc wireless networks.*
- *The Collision Avoidance Time Allocation (CATA) is based on dynamic topology-dependent transmission scheduling.*
- *Five-Phase Reservation Protocol (FPRP) is a single-channel time division multiple access based broadcast scheduling protocol.*

Chapter III

Overview of Basic Routing Protocols

Objective

- *To learn about the fundamentals of Routing Protocol*
- *To know the classifications of Routing protocol*
- *To become familiar with the concepts of DSR and AODV protocols*
- *To understand the application of DSDV*

3.0. Routing Protocol

- Routing protocols adapt to changing network conditions and by definition offer multihop paths. Routing protocols differ in route table
    - construction
    - maintenance
    - update
- Next-hop routing protocols can be categorized as:
    - Link-state
    - Distance-vector
- Proactive vs. Reactive Routing Protocols
    - Proactive Protocols
        - Have lower latency due to maintenance of routes at all times.
        - Can result in much higher overhead due to frequent route updates.
    - Reactive Protocols may have

    - Higher latency since the routes have to be discovered when the source node initiates a route request.
    - Lower overhead since routes are maintained only on-demand basis.
- Hybrid protocols
    - Adaptive; Combination of proactive and reactive.
    - Example: ZRP (zone routing protocol).

3.1. PROTOCOLS

- DSR: Dynamic Source Routing:
- ABR: Associativity-Based Routing
- SSA: Signal Stability-Based Adaptive Routing Algorithm
- AODV: Ad Hoc On-Demand Distance Vector
- LAR: Location Aided Routing Protocol
- RDMAR: Relative Distance Micro-Discovery Ad Hoc Routing
- LMR: Light-weight Mobile Routing
- TORA: Temporally Ordered Routing Algorithm
- ARA: Ant-colony-based Routing Algorithm

3.1.1. Dynamic Source Routing (DSR)

- **Dynamic Source Routing (DSR)** is a routing protocol for wireless mesh networks. It is similar to AODV in that it forms a route on-demand when a transmitting node requests one.
- However, it uses source routing instead of relying on the routing table at each intermediate device. Determining source routes requires accumulating the address of each device between the source and destination during route discovery.
- The accumulated path information is cached by nodes processing the route discovery packets. The learned paths are used to route packets. To accomplish source routing, the routed packets contain the address of each device the packet will traverse.
- This may result in high overhead for long paths or large addresses, like IPv6. To avoid using source routing, DSR optionally defines a

flow id option that allows packets to be forwarded on a hop-by-hop basis.

- This protocol is truly based on source routing whereby all the routing information is maintained (continually updated) at mobile nodes. It has only two major phases, which are Route Discovery and Route Maintenance. Route Reply would only be generated if the message has reached the intended destination node (route record which is initially contained in Route Request would be inserted into the Route Reply).
- To return the Route Reply, the destination node must have a route to the source node. If the route is in the Destination Node's route cache, the route would be used. Otherwise, the node will reverse the route based on the route record in the Route Request message header (this requires that all links are symmetric).
- In the event of fatal transmission, the Route Maintenance Phase is initiated whereby the Route Error packets are generated at a node. The erroneous hop will be removed from the node's route cache; all routes containing the hop are truncated at that point. Again, the Route Discovery Phase is initiated to determine the most viable route.
- Dynamic source routing protocol (DSR) is an on-demand protocol designed to restrict the bandwidth consumed by control packets in ad hoc wireless networks by eliminating the periodic table-update messages required in the table-driven approach.
- The major difference between this and the other on-demand routing protocols is that it is beacon-less and hence does not require periodic hello packet (beacon) transmissions, which are used by a node to inform its neighbors of its presence. The basic approach of this protocol (and all other on-demand routing protocols) during the route construction phase is to establish a route by flooding RouteRequest packets in the network.
- The destination node, on receiving a RouteRequest packet, responds by sending a RouteReply packet back to the source, which carries the route traversed by the RouteRequest packet received.

Advantages and Disadvantages

- This protocol uses a reactive approach which eliminates the need to periodically flood the network with table update messages which are required in a table-driven approach.

- In a reactive (on-demand) approach such as this, a route is established only when it is required and hence the need to find routes to all other nodes in the network as required by the table-driven approach is eliminated.
- The intermediate nodes also utilize the route cache information efficiently to reduce the control overhead. The disadvantage of this protocol is that the route maintenance mechanism does not locally repair a broken link. Stale route cache information could also result in inconsistencies during the route reconstruction phase.
- The connection setup delay is higher than in table-driven protocols. Even though the protocol performs well in static and low-mobility environments, the performance degrades rapidly with increasing mobility. Also, considerable routing overhead is involved due to the source-routing mechanism employed in DSR.
- This routing overhead is directly proportional to the path length.

3.1.2. DSDV Routing

- Destination-Sequenced Distance Vector (DSDV) routing protocol is a pro-active, table-driven routing protocol for MANETs developed by Charles E. Perkins and Pravin Bhagwat in 1994. It uses the hop count as metric in route selection.
- DSDV Routing Table: Every node will maintain a table listing all the other nodes it has known either directly or through some neighbors. Every node has a single entry in the routing table.
- The entry will have information about the node's IP address, last known sequence number and the hop count to reach that node. Along with these details the table also keeps track of the nexthop neighbor to reach the destination node, the timestamp of the last update received for that node.
- The DSDV update message consists of three fields, Destination Address, Sequence Number and Hop Count.
- Each node uses 2 mechanisms to send out the DSDV updates.
- They are

  Periodic Updates:

  Periodic updates are sent out after every m_periodic UpdateInterval(default:15s). In this update the node broadcasts out its entire routing table.

Trigger Updates:
Trigger Updates are small updates in-between the periodic updates. These updates are sent out whenever a node receives a DSDV packet that caused a change in its routing table. The original paper did not clearly mention when for what change in the table should a DSDV update be sent out. The current implementation sends out an update irrespective of the change in the routing table.

- The updates are accepted based on the metric for a particular node. The first factor determining the acceptance of an update is the sequence number. It has to accept the update if the sequence number of the update message is higher irrespective of the metric. If the update with same sequence number is received, then the update with least metric (hopCount) is given precedence.
- In highly mobile scenarios, there is a high chance of route fluctuations, thus the concept of weighted settling time where an update with change in metric will not be advertised to neighbors. The node waits for the settling time to make sure that it did not receive the update from its old neighbor before sending out that update.
- The current implementation covers all the above features of DSDV. The current implementation also has a request queue to buffer packets that have no routes to destination. The default is set to buffer up to 5 packets per destination.

Destination-Sequenced Distance-Vector Routing (DSDV) is a table-driven routing scheme for ad hoc mobile networks based on the Bellman–Ford algorithm. It was developed by C. Perkins and P. Bhagwat in 1994. The main contribution of the algorithm was to solve the routing loop problem.

Each entry in the routing table contains a sequence number, the sequence numbers are generally even if a link is present; else, an odd number is used. The number is generated by the destination, and the emitter needs to send out the next update with this number. Routing information is distributed between nodes by sending full dumps infrequently and smaller incremental updates more frequently.

Advantages

- The availability of paths to all destinations in network always shows that less delay is required in the path set up process.

- The method of incremental update with sequence number labels, marks the existing wired network protocols adaptable to Ad-hoc wireless networks. Therefore, all available wired network protocol can be useful to ad hoc wireless networks with less modification.

Disadvantages

- DSDV requires a regular update of its routing tables, which uses up battery power and a small amount of bandwidth even when the network is idle.
- Whenever the topology of the network changes, a new sequence number is necessary before the network re-converges; thus, DSDV is not suitable for highly dynamic or large scale networks. (As in all distance-vector protocols, this does not perturb traffic in regions of the network that are not concerned by the topology change.)

Influence

- While DSDV itself does not appear to be much used todayother protocols have used similar techniques. The best-known sequenced distance vector protocol is AODV, which, by virtue of being a reactive protocol, can use simpler sequencing heuristics. Babel is an attempt at making DSDV more robust, more efficient and more widely applicable while staying within the framework of proactive protocols.

3.1.3. AODV (Ad hoc On-Demand Distance Vector)

Introduction

AODV enables "dynamic, self-starting, multi-hop routing between mobile nodes wishing to establish and maintain an ad hoc network. AODV allows for the construction of routes to specific destinations and does not require that nodes keep these routes when they are not in active communication. AODV avoids the "counting to infinity" problem by using destination sequence numbers. This makes AODV loop-free.

Overview

AODV defines 3 message types:

- Route Requests (RREQs)
  - RREQ messages are used to initiate the route finding process.
- Route Replies (RREPs)
  - RREP messages are used to finalize the routes.
- Route Errors (RERRs)
  - RERR messages are used to notify the network of a link breakage in an active route.

The AODV protocol is only used when two endpoints do not have a valid active route to each other. Nodes keep a "precursor list" that contains the IP address for each of its neighbours that are likely to use it for a next hop in their routing table. Route table information must be kept for all routes even short-lived routes. The routing table fields used by AODV are:

- Destination IP Address
- Destination Sequence Number
- Valid Destination Sequence number flag
- Other state and routing flags
- Network Interface
- Hop Count
- Next Hop
- List of Precursors
- Lifetime

AODV Terminology

This protocol specification uses conventional meanings for capitalized words such as MUST, SHOULD, etc., to indicate requirement levels for various protocol features. This section defines other terminology used with AODV.

- ACTIVE ROUTE:
  - A route towards a destination that has a routing table entry that is marked as valid. Only active routes can be used to forward data packets.

- BROADCAST:
  - Broadcasting means transmitting to the IP Limited Broadcast address, 255.255.255.255. A broadcast packet may not be blindly forwarded, but broadcasting is useful to enable dissemination of AODV messages throughout the ad hoc network.
- DESTINATION:
  - An IP address to which data packets are to be transmitted. Same as "destination node". A node knows it is the destination node for a typical data packet when its address appears in the appropriate field of the IP header. Routes for destination nodes are supplied by action of the AODV protocol, which carries the IP address of the desired destination node in route discovery messages.
- FORWARDING NODE:
  - A node that agrees to forward packets destined for another node, by retransmitting them to a next hop that is closer to the unicast destination along a path that has been set up using routing control messages.
- FORWARD ROUTE
  - A route set up to send data packets from a node originating a Route Discovery operation towards its desired destination.
- INVALID ROUTE
  - A route that has expired, denoted by a state of invalid in the routing table entry. An invalid route is used to store previously valid route information for an extended period of time. An invalid route cannot be used to forward data packets, but it can provide information useful for route repairs, and also for future RREQ messages.
- ORIGINATING NODE
  - A node that initiates an AODV route discovery message to be processed and possibly retransmitted by other nodes in the adhoc network. For instance, the node initiating a Route Discovery process and broadcasting the RREQ message is called the originating node of the RREQ message.
- REVERSE ROUTE
  - A route set up to forward a reply (RREP) packet back to the originator from the destination or from an intermediate node having a route to the destination.

- SEQUENCE NUMBER
  - A monotonically increasing number maintained by each originating node. In AODV routing protocol messages, it is used by other nodes to determine the freshness of the information contained from the originating node.

Applicability Statement

The AODV routing protocol is designed for mobile ad hoc networks with populations of tens to thousands of mobile nodes. AODV can handle low, moderate, and relatively high mobility rates, as well as a variety of data traffic levels. AODV is designed for use in networks where the nodes can all trust each other, either by use of preconfigured keys, or because it is known that there are no malicious intruder nodes. AODV has been designed to reduce the dissemination of control traffic and eliminate overhead on data traffic, in order to improve scalability and performance.

3.1.4. WRP (Wireless Routing Protocol)

WRP uses an enhanced version of the distance-vector routing protocol, which uses the Bellman–Ford algorithm to calculate paths. Because of the mobile nature of the nodes within the MANET, the protocol introduces mechanisms which reduce route loops and ensure reliable message exchange.

- WRP, similar to Destination-Sequenced Distance Vector routing (DSDV), inherits the properties of the distributed Bellman–Ford algorithm.
- To counter the count-to-infinity problem and to enable faster convergence, it employs a unique method of maintaining information regarding the shortest distance to every destination node in the network and the penultimate hop node on the path to every destination node.
- Since WRP, like DSDV, maintains an up-to-date view of the network, every node has a readily available route to every destination node in the network.
- It differs from DSDV in table maintenance and in the update procedures. While DSDV maintains only one topology table, WRP uses a set of tables to maintain more accurate information.
- The tables that are maintained by a node are the following:

- distance table (DT)
- routing table (RT)
- link cost table (LCT)
- message retransmission list (MRL).

- The DT contains the network view of the neighbours of a node. It contains a matrix where each element contains the distance and the penultimate node reported by a neighbour for a particular destination. The RT contains the up-to-date view of the network for all known destinations.
- It keeps the shortest distance, the predecessor node (penultimate node), the successor node (the next node to reach the destination), and a flag indicating the status of the path.
- The path status may be a simple path (correct), or a loop (error), or the destination node not marked (null). The LCT contains the cost (e.g., the number of hops to reach the destination) of relaying messages through each link.
- The cost of a broken link is infinity. It also contains the number of update periods (intervals between two successive periodic updates) passed since the last successful update was received from that link.
- This is done to detect links breaks. The MRL contains an entry for every update message that is to be retransmitted and maintains a counter for each entry.
- This counter is decremented after every retransmission of an update message. Each update message contains a list of updates. A node also marks each node in the RT that has to acknowledge the update message it transmitted.
- Once the counter reaches zero, the entries in the update message for which no acknowledgments have been received are to be retransmitted and the update message is deleted.
- Thus, a node detects a link break by the number of update periods missed since the last successful transmission. After receiving an update message, a node not only updates the distance for transmission neighbours but also checks the other neighbours distance, hence convergence is much faster than DSDV.

Cluster Gateway Switch Routing (CGSR)

- Nodes are grouped into clusters with one cluster head in charge of nodes in a cluster.

- Distributed cluster head selection algorithm is used to elect cluster heads.
- This algorithm needs to be invoked when a cluster head moves away.
- Least Cluster Change (LCC) algorithm is introduced to avoid re-invoking election algorithm when the cluster membership update occurs.
- LCC algorithm allows the invocation of election algorithm when two cluster head come into contact or when a node no longer be able to attach to any cluster heads.
- Uses DSDV as an underlying routing scheme.
- The route traffic follows:
- Source → Cluster head → Gateway → Cluster head …→ Destination.
- Each node keeps a cluster member table which stores the destination cluster head for each node in the network.
- Every node periodically broadcasts these cluster member tables using DSDV protocol.
- Nodes update their cluster member tables upon receiving these broadcasts.
- Each node also maintains a routing table that is used to determine the next hop towards the destination.
- When a packet is received, the node checks both cluster member and routing tables to determine the nearest cluster head in the route to the destination.
- The node checks its routing table to determine the next hop node to reach cluster head.
- Both cluster member and routing tables need to be updated.

Advantages

1. Better bandwidth Utilization.
2. Easy implementation of Priority scheduling with token scheduling and gateway scheduling.

Disadvantages

1. Gateway conflicts will occur. In order to avoid these conflicts more resources are needed.

2. Power consumption of the cluster head is more when compared to normal node.

3.1.5. ABR

- Free from loops, deadlock, and packet duplicates, ABR defines a new routing metric for ad-hoc mobile networks
- Each node generates periodic beacons (hello messages) to signify its existence to the neighbors
- These beacons are used to update the associativity table of each node
- With the temporal stability and the associativity table the nodes are able to classify each neighbor link as stable or unstable
  - Select Longer-Lived routes.
- Beacon based protocol.
- Defining the Location Stability between nodes
  1. Used as a metric instead of shortest hop.
  2. Determined by beacon counting.
- Links between nodes classified into Stable and Unstable link according to beacon counts.

ABR Terminology

- Use some words which have some specific meaning in the context of ABR.
- ABR - Associativity Based Routing using the principle of associativity.
- MH - Mobile host.
- SRC - Mobile host which desires a route.
- DEST - Mobile host to which information sent by SRC will be received.
- IN - Intermediate nodes between SRC and DEST.
- BQ - Broadcast query packet used by SRC when it requires a new route.
- REPLY - This message is sent by the DEST node in response to a BQ.
- SEQ No- It is used to uniquely identify each BQ packet, so that no BQ packet will be broadcast more than once.

- LQ - Localized Query control packet which is used during route reconstruction.
- RRC - Route Reconstruction Phase.
- RN - Route Notification control packet.
- RD - Route Deletion control packet

ABR– Route Discovery

- Source Node broadcasts Route Request packet.
- Each intermediate node do the following steps:
- If request received before → discard.
- If node ID is listed in request → discard.
- If route to the destination is available → send Route Reply.
- Otherwise → append node ID and Beacon Count and rebroadcast.
- Destination node
- Once getting the first Route Request, it waits for certain period to receive multiple Route Requests
- From multiple routes, it selects the route with maximum proportion of stable links
- If more than one route has the maximum proportion of stable links, the shortest path is selected
- Only single route is selected by the destination

ABR Concern

- Chosen path may not be the shortest path
- May lead to higher delay in route repairs
- Single path selection
- High cost of periodic beaconing
- Power
- Bandwidth

ABR Consists of 3 Phases

- Route Discovery
- Route Repair/Reconstruction
- Route Delete

Advantages

- Free from duplicate packets.
- Stable routes have higher preference when compared to shorter routes.
- They result in fewer path breaks which in turn reduce the extent of flooding due to reconfiguration of paths in the network.

Disadvantages

- Short beaconing interval to reflect association degree precisely.
- This protocol's chosen path may be longer than the shortest path between the source and destination because of the preference given to stable paths.
- The repetitive LQ broadcasts may result in high delays during route repairs.

3.1.6. Source Tree Adaptive Routing Protocol (STAR)

The source tree Adaptive Routing Protocol (STAR) was the first proactive routing protocol, which worked with link state information. In addition, it implemented for the first time the LORA principle (Least overhead routing approach; getting found paths as long as possible in order to avoid control messages). STAR does not use the shortest practicable paths, so that unnecessary control messages are avoided. There are all the nodes provided with fixed addresses, which has the advantage that no constantly new updates of information are needed. These consist of at least one LSU (Link State Update).

Update Information

Update information as broadcast is shipped and numbered. In this case, the counter is incremented only by the transmitter. As LSU is valid if the number is higher than the number for the same connection last saved, and this has the advantage that LSUs need not to be updated periodically.

Updated information is sent, if

- a receiver is no longer attainable,
- A new recipient is found

- If it appears that loops are formed,
- the metric of connections exceeds the maximum value, which can be determined by comparing the received with the own source tree.

Preventing Loops

To prevent loops, the following rules for sending updated information must apply to a router. It sends updates if

1. a path in a loop end
2. selected a new successor to the router, which has a higher address,
3. the distance to the receiver over a selected successor is greater than that of the old successor.

Star Overview

In STAR, each router reports to its neighbors the characteristics of every link it uses to reach a destination. The set of links used by a router in its preferred path to destinations is called the source tree of the router. A router knows its adjacent links and also the source trees reported by its neighbors; the aggregation of a router's adjacent links and the source trees reported by its neighbors constitute a partial topology graph. The links in the source tree and topology graph must be adjacent links or links reported by at least one neighbor. The router uses the topology graph to generate its own source tree. Each router derives a routing table specifying the successor to each destination by running a local route-selection algorithm on its source tree.

Under ORA, updates are sent when source trees change. Under LORA, a router running STAR sends updates on its source tree to its neighbors only when it loses all paths to one ore more destinations, when it detects a new destination, or when it determines that local changes to its source tree can potentially create long term routing loops. Because each router communicates its source tree to its neighbors, the deletion of a link is no longer used to reach a destination. This is implicit with the addition of the new link used to reach the destination and need not be sent explicitly as an update; a router makes explicit reference to a failed link only when the deletion of a link causes the router to have no paths to one or more destinations, in which case the router cannot provide new links to make the deletion of the failed link implicit.

The basic update unit used in STAR to communicate changes to source trees is the link-state update (LSU). An LSU reports the characteristics of a link; an update message contains one or more LSUs. For a link between router u and router or destination v, router u is called the headnode of the link in the direction from u to v. The head node of a link is the only router that can report changes in the parameters of that link. LSUs are validated using sequence numbers, and each router erases a link from its topology graph if the link is not present in the source trees of any of its neighbors. The head of a link does not periodically send LSUs for the link, because link-state information never ages out.

Unlike any of the hierarchical link-state routing schemes proposed to date for packet-radio networks, STAR does not require backbones, the dissemination of complete cluster topology within a cluster, or the dissemination of the complete inter-cluster connectivity among clusters. Furthermore, STAR can be used with distributed hierarchical routing schemes proposed in the past for both distance-vector or link-state routing.

Prior proposals for link-state routing using partial link-state data without clusters require routers to explicitly inform their neighbors which links they use and which links they stop using. In contrast, because STAR sends only changes to the structure of source trees, and because each destination has a single predecessor in a source tree, a router needs to send only updates for those links that are part of the tree and a single update entry for the root of any subtree of the source tree that becomes unreachable due to failures. Routers receiving a STAR update can infer correctly all the links that the sender has stopped using, without the need for explicit delete updates.

Advantages and Disadvantages

STAR has very low communication overhead among all the table-driven routing protocols

The use of LORA approach in these table-driven routing protocols reduces the average control overhead compared to several other on-demand routing protocols.

3.1.7. Signal Stability-Based Adaptive Routing Protocol (SSR)

Signal Stability-based Adaptive Routing Protocol (SSA) tried more routes based on signal strength and stability of the local node. SSA is partly based on DSR. The main objective of SSA is that only routes are chosen that have stronger ties.

SSA is just like the Associativity Based Routing Protocol (ABR) Beacon-based, which means that the signal strength of the beacons (special messages) is measured to assess the connection stability between nodes. The local stability defined paths already exist for a long time. With the aid of connection stability and local stability the compounds are classified as stable or unstable.

The signal strength of the neighboring nodes is stored in a table called SST. Each node constantly renews its SST using the received beacon packets. The information from the SST be used to forward the route request message to the destination, but only for stable compounds.

One difference between SSA and ABR is that node route request message is only forwarded when they have this condition with a stable connection. Otherwise, the corresponding packets are discarded. Furthermore, SSA uses only one metric: the signal strength between nodes.

1. First, the network with route request packets are flooded.
2. Now, if a node receives such a packet, it checks whether the packet was sent on a stable connection. All packets that it received twice over an unstable connection are discarded. In Route-Request packet, the address of each intermediate node of the previous route will be saved (thus this modified before each tell).
3. If the first route request packet to the target via a classified a stable connection arrives, the destination immediately sends a route reply packet to the source to inform the selected route. The route reply packet is sent back through the same path.
4. Before SSA is used, the initiator is checked, whether in the Route-cache a route to the destination exists. Only if no such route exists, SSA is used.

Maintaining a Route

1. Immediately when a connection is interrupted, the two intermediate nodes affected send a corresponding message to the source node and the destination node.
2. The start node floods the network again with route request messages to find another stable route.
3. Older Routes are only deleted if the data packet with the route information does not reach the next node.
4. There is no way stable compounds can be found, and unstable connections are used.

Signal Stability-Based Adaptive Routing protocol (SSR) presented in an on-demand routing protocol that routes based on the signal strength between nodes and a node's location stability. This route selection criterion has the effect of choosing routes that have "stronger" connectivity. SSR comprises of two cooperative protocols: the Dynamic Routing Protocol (DRP) and the Static Routing Protocol (SRP).

The DRP maintains the Signal Stability Table (SST) and Routing Table (RT). The SST stores the signal strength of neighboring nodes obtained by periodic beacons from the link layer of each neighboring node. Signal strength is either recorded as a strong or weak channel. All transmissions are received by DRP and processed. After updating the appropriate table entries, the DRP passes the packet to the SRP.

The SRP passes the packet up the stack if it is the intended receiver. If not, it looks up the destination in the RT and forwards the packet. If there is no entry for the destination in the RT, it initiates a route-search process to find a route. Route-request packets are forwarded to the next hop only if they are received over strong channels and have not been previously processed (to avoid looping). The destination chooses the first arriving route-search packet to send it back as it is highly likely that the packet arrived over the shortest and/or least congested path. The DRP reverses the selected route and sends a route-reply message back to the initiator of route-request. The DRP of the nodes along the path update their RTs accordingly.

Route-search packets arriving at the destination have necessarily arrived on the path of strongest signal stability because the packets arriving over a weak channel are dropped at the intermediate nodes. If the source times out before receiving a reply then it changes the PREF field in the header to indicate that weak channels are acceptable, since these may be the only links over which the packet can be propagated.

When a link failure is detected within the network, the intermediate nodes send an error message to the source indicating which channel has failed. The source then sends an erase message to notify all nodes of the broken link and initiates a new route-search process to find a new path to the destination.

Principles of SSA

To select routes based on the signal strength between nodes and on a node's location stability. To choose routes that has stronger connectivity. SSA has two component co-operative protocols:

1. The Dynamic Routing Protocol
2. The Static Routing Protocol

The Dynamic Routing Protocol (DRP)

- The DRP is responsible for maintaining what is called the Signal Stability Table and also the Routing Table.
- SST – record of signal strengths of neighboring nodes which is obtained by means of periodic beaconing.
- Quantized levels possible → weak channel vs. strong channel.
- When a packet is received, DRP processes the packet, updates the tables and passes the received packet to the SRP.

The Static Routing Protocol

- Forwards the packet up to the transport layer if it is the receiver.
- If not, it looks up the routing table and forwards the packet to the appropriate next-hop.
- If no entry is found, it initiates a route search

Advantages

The main advantage of SSA is that this protocol finds more stable routes to its destination place as DSR. The shortest path is not necessarily the best. By checking the signal strength via beacons SSA can find a stable route.

Disadvantages

Aborted connections are discovered locally, but not repaired. Multiple flooding of the network with route request messages restricts the bandwidth of the network. In addition, Route-Request packets are not considered over weak links but directly rejected.

3.1.8. Optimized Link State Routing Protocol

- The **Optimized Link State Routing Protocol (OLSR)** is an IP routing protocol optimized for mobile ad hoc networks, which can also be used on other wireless ad hoc networks.

- OLSR is a proactive link-state routing protocol, which uses hello and topology control (TC) messages to discover and then disseminate link state information throughout the mobile ad hoc network.
- Individual nodes use this topology information to compute next hop destinations for all nodes in the network using shortest hop forwarding paths.
- Link-state routing protocols such as Open Shortest Path First (OSPF) and IS-IS elect a designated router on every link to perform flooding of topology information.
- In wireless ad hoc networks, there is a different notion of a link, packets can and do go out the same interface; hence, a different approach is needed in order to optimize the flooding process.
- Using Hello messages the OLSR protocol at each node discovers 2-hop neighbor information and performs a distributed election of a set of multipoint relays (MPRs). Nodes select MPRs such that there exists a path to each of its 2-hop neighbors via a node selected as an MPR.
- These MPR nodes then source and forward TC messages that contain the MPR selectors.
- This functioning of MPRs makes OLSR unique from other link state routing protocols in a few different ways: The forwarding path for TC messages is not shared among all nodes but varies depending on the source, only a subset of nodes source link state information, not all links of a node are advertised but only those that represent MPR selections.
- Since link-state routing requires the topology database to be synchronized across the network, OSPF and IS-IS perform topology flooding using a reliable algorithm.
- Such an algorithm is very difficult to design for ad hoc wireless networks, so OLSR doesn't bother with reliability.
- It simply floods topology data often enough to make sure that the database does not remain unsynchronized for extended periods of time.

Benefits

- Being a proactive protocol, routes to all destinations within the network are known and maintained before use.

- Having the routes available within the standard routing table can be useful for some systems and network applications as there is no route discovery delay associated with finding a new route.
- The routing overhead generated, while generally greater than that of a reactive protocol, does not increase with the number of routes being created.
- Default and network routes can be injected into the system by HNA messages allowing for connection to the internet or other networks within the OLSR MANET cloud.
- Network routes are something reactive protocols do not currently execute well.
- Timeout values and validity information is contained within the messages conveying information allowing for differing timer values to be used at differing nodes.
- The original definition of OLSR does not include any provisions for sensing of link quality; it simply assumes that a link is up if a number of hello packets have been received recently.
- This assumes that links are bi-modal (either working or failed), which is not necessarily the case on wireless networks, where links often exhibit intermediate rates of packet loss. Implementations such as the open source OLSRd (commonly used on Linux-based mesh routers) have been extended (as of v. 0.4.8) with link quality sensing.
- Being a proactive protocol, OLSR uses power and network resources in order to propagate data about possibly unused routes.
- While this is not a problem for wired access points, and laptops, it makes OLSR unsuitable for sensor networks that try to sleep most of the time.
- For small scale wired access points with low CPU power, the open source OLSRd project showed that large scale mesh networks can run with OLSRd on thousands of nodes with very little CPU power on 200 MHz embedded devices.
- Being a link-state protocol, OLSR requires a reasonably large amount of bandwidth and CPU power to compute optimal paths in the network.
- In the typical networks where OLSR is used (which rarely exceed a few hundreds of nodes), this does not appear to be a problem.
- By only using MPRs to flood topology information, OLSR removes some of the redundancy of the flooding process, which may be a problem in networks with moderate to large packet loss rates –

however the MPR mechanism is self-pruning (which means that in case of packet losses, some nodes that would not have retransmitted a packet, may do so).

- OLSR makes use of "Hello" messages to find its one hop neighbors and its two hop neighbors through their responses.
- The sender can then select its multipoint relays (MPR) based on the one hop node that offers the best routes to the two hop nodes.
- Each node also has an MPR selector set, which enumerates nodes that have selected it as an MPR node.
- OLSR uses topology control (TC) messages along with MPR forwarding to disseminate neighbor information throughout the network.
- Host and network association (HNA) messages are used by OLSR to disseminate network route advertisements in the same way TC messages advertise host routes.
- The problem of routing in ad hoc wireless networks is actively being researched, and OLSR is but one of several proposed solutions.
- To many, it is not clear whether a whole new protocol is needed, or whether OSPF could be extended with support for wireless interfaces.
- In bandwidth- and power-starved environments, it is interesting to keep the network silent when there is no traffic to be routed.
- Reactive routing protocols do not maintain routes, but build them on demand.
- As link-state protocols require database synchronisation, such protocols typically use the distance vector approach, as in AODV and DSDV, or more ad hoc approaches that do not necessarily build optimal paths, such as Dynamic Source Routing.

3.1.9. Flow-Oriented Routing Protocols (AD-HOC Networks)

General Characteristics

Finds a route on demand by following existing flows. For example, one option is to unicast consecutively when forwarding data while promoting a new link.

Lightweight Mobile Routing (LMR)

The LMR agorithm is a link reversal algorithm which addresses the issue of partitioned networks found with the Gafni-Bertsekas algorithms, by providing a link erasure mechanism.

Pros

- In a rapidly changing network, routes will be found rather quickly and broken links will have only local affect.
- One can expect good results if the network connectivity is rather high (dense network).
- A higher level protocol could use redundant routes in a round-robin fashion to economically use local bandwidth.

Cons

- In a rapidly changing network there may be many false RPY packets producing message overhead.
- Routes may be redundant.
  - The key feature of this protocol is applying a prediction based scheme for selecting and maintaining its routes.
  - It can predict the link expiration time (LET) for a given link-for a complete description about used prediction algorithm, and consequently it can predict a route expiration time (RET) for a given path.
  - FORP uses such predictions to select the longest likely to live paths and to handoff the current sessions and find alternative paths before the expiration of the currently used ones.
  - As explained, this scheme allows making routing decisions which ensure some level of quality of service (QoS).
  - Additionally, the performance study shows less control overhead required by this class of prediction based protocols.
  - However, a common timing reference between the nodes in the networks is required which again introduces the complexity of depending on another resource, namely, GPS.
  - When a source node needs to send packets to a destination node, first, it checks its own routing table.

- If it already has an unexpired path to the destination, it sends the packet directly to the destination.
- If not, it initiates a route request packet which carries a flow identification number and a sequence number along with the source node address and the destination node address.
- Upon receiving the route request message, an intermediate node checks the sequence number in the message, and then it discards the packet if the sequence number is less than the last received sequence number associated with the flow identification number from this source.
- In case of equal sequence numbers, the intermediate node forwards the route request message only if it has received it form a path of a larger RET.
- Otherwise, the intermediate node adds the LET of the link it received the message from, and adds its address and then broadcasts the packet.
- This way the route request message arrives to the destination node continuing the entire traversed path along with its RET.
- A path is used instead of a currently used one in case it has a longer RET. For the route maintenance process, FORP defines a critical as the difference between the RET of the currently used path and the time the latest packet took to traverse along the path.
- This time is also affected by the continuously received RET values form the intermediate nodes along with the data packets.
- When the destination node detects that the critical time is about to be reached, it sends a route hand off packet to the source route which initiates the route setup process again based on the RET values, the number of hops or any other information included in the route handoff packet.
- A large number of on-demand routing protocols have already been proposed.
- Each protocol has its own key features, which may add positive or negative sides to the protocol.
- However, on-demand routing protocols share their common ability to adopt with the dynamically changing topology of the wireless ad hoc networks, in spite of the delay required to find routes for destination nodes.

3.1.10. FSR Protocol

Fisheye State Routing (FSR) is an implicit hierarchical routing protocol. Also considered a proactive protocol it is a link state based routing protocol that has been adapted to the wireless ad hoc environment. It relays on link state protocol as a base, and has the ability to provide route information instantly by maintaining a topology map at each node. Thus the FSR protocol will maintain updated information from the neighbor node through a link state table.

According to Kleinrock and Stevens, FSR uses the "fisheye" technique which was used to reduce the size of information required to represent graphical data. The eye of a fish captures with high detail the pixels near the focal point. The detail decreases as the distance from the focal point increases.

In routing, the fisheye approach translates to maintaining accurate distance and path quality information about the immediate neighborhood of a node, with progressively less details as the distance increases.

Algorithm

FSR is based on a link-state foundation updating mechanism, which has the feature of maintaining a topology map at each node. This mechanism reduces the control overhead by disseminating topology information using the fisheye technique, where routing information is updated at different rates depending on the distance from the source. And it can be broken down into

- A Node stores the Link State for every destination in the network
- A Node periodically broadcasts update messages to its neighbors
- Updates correspond to closer nodes propagating more frequently

Tasks

In this routing protocol, there are three major tasks

1. **Neighbor Discovery**: responsible for establishing and maintaining neighbor relationships.

2. **Information Dissemination**: responsible for disseminating Link State Packets (LSP), which contain neighbor link information, to other nodes in the network.
3. **Route Computation**: responsible for computing routes to each destination using the information of the LSPs.

Initially every node start has an empty topology table and an empty neighbor list. Invoking the Neighbor discovery mechanism in order to acquire neighbors and to maintain current neighbor relationships after its local variables are initialized. By using the Information Dissemination mechanism, the distribution of LSP in the network is produced. Each node has a database consisting of the collection of LSPs originated by each node in the network. From this database, the node uses the Route Computation mechanism to yield a routing table for the protocol. This process is periodically repeated.

Applications

One use of the FSR is reducing overhead control traffic. It has also shown a good performance in terms of successful packet delivery in the presence of low mobility. During the case of high mobility; in order to get a successful packet delivery, the update interval time must be properly selected

Algorithms Related to FSR Protocol

A similar algorithm to FSR is the Link-state routing protocol. It has a different approach in which routing information updates entries in routing. FSR uses the fisheye technique which allows considering a fraction of the scope of the eye, leading to reduction in the message size. Thus routing update overhead is reduced.

Advantages and Disadvantages

The notion of multiple level scopes employed by FSR significantly reduces the bandwidth consumed by link state update packets. Hence, FSR is suitable for large and highly mobile ad hoc wireless networks. The choice of the number of hops associated with each scope level has a significant influence on the performance of the protocol at different mobility values, and hence must be carefully chosen.

PLBR Protocol

Routing protocols in wireless ad hoc networks experience high control overhead due to frequent path breaks that occur due to mobility of nodes, which leads to flooding of control packets throughout the network. The preferred link based routing protocol reduces flooding of control packets by selectively allowing some nodes to forward the packets using a preferred list.

Wireless mobile networks provide ubiquitous connectivity for the paradigm of any-time-any-where access. These wireless networks are classified as those having a fixed infrastructure, like cellular networks and those without infrastructure, referred to as wireless ad hoc networks. Cellular networks divide the area to be served into cells for spatial reuse of scarce bandwidth. A fixed base-station serves all mobile nodes in a cell and connects them to other fixed/mobile networks. Wireless ad hoc networks are infrastructure-less networks with no fixed stations. All nodes participate in maintaining connectivity and act as routers/switches as well as endpoints. If communicating nodes are separated by a distance larger than their transmission range, connectivity is provided using multiple wireless links. These networks can be deployed quickly and are used for collaborative computing, search and rescue operations, military applications and emergency situations like floods/earthquakes. Designing a routing protocol for ad hoc networks is a challenging task due to issues such as mobility of nodes, limited bandwidth, error-prone wireless links, shared broadcast radio links, and the hidden terminal problem.

Mechanism

Preferred links are not explicitly used by any routing protocol in ad hoc networks during forwarding of route requests. But SSA does it in an implicit way by processing RReq only if received through strong links. Wired networks also employ preferred link mechanisms, but restrict themselves by selecting a single preferred link, based on heuristics that satisfy multiple constraints, i.e., minimum cost and least delay required by route. These algorithms, in case RReq fail, employ backtracking to trace all possibilities.

The heuristics can be destination based or session based. Destination based heuristics share the same preferred link for the same destination for different RReq, while session based heuristics choose the preferred link based on per session requirements of RReq packets. In ad hoc networks, the single preferred link model is not suitable due to reasons like lack of topology information, continuously changing link characteristics, broadcast radio channel, and mobility of nodes. Moreover, backtracking will cause longer path lengths and higher delay.

Preferred List = Subset of the Neighbors List

- Reactive routing protocols
- Basic concept:
  - Each node maintains two tables: NT and NNT
  - Each node selects a subset called Preferred List (PL)
  - K: the size of the PL
- Preferred List construction:
  - Neighbor Degree-based Preferred Link Algorithm
    - based on neighbor nodes' degree
    - divides its neighbor nodes to reachable and unreachable
  - Weight-based Preferred Link algorithm
    - based on the weight given to a node
    - its weight is based on its neighbors' temporal and spatial stability

The Preferred Link-based Routing (PLBR) is a reactive routing protocol which applies particular mechanisms to avoid flooding the network with RouteRequest packets. It provides a mechanism where not all the nodes are allowed to forward the received RouteRequest packet. Hence, it reduces the control overhead. Here, every node maintains a table with its neighbors (NT) and the neighbor's neighbors (NNT). A node selects always a subset of nodes from its neighbor list (NL). This subset is called the preferred list (PL).

The RouteRequest packet contains the PL of the sender and is broadcasted to its neighbors. The ones that are in the PL will forward the RouteRequest packet. The others will discard it. The size of the PL is K, where K is the maximum number of nodes comprising the PL. PL is always a subset of PLT where PLT is the preferred list table maintained by every node and which contains all neighbor's nodes in order of preference. Any change in neighbors' topology is maintained by sending a beacon that updates the tables. PLBR consists of three different

Phases:

- Route establishment
- Route selection
- Route maintenance.

During the first phase of route establishment, the source node s is trying to establish a route to the destination node d. If node d is in source's NNT, the route is established directly and the information is forwarded to the destination.

The source initiates a Route Request packet which contains:

- **Source node's address:** SourceID
- **Destination node's address:** DestinationID
- **Unique sequence number:** SeqNum, which prevents the forwarding of multiple copies of the same RouteRequest packet received from different neighbors.
- **Traversed Path:** TP, which indicates the nodes that have been visited so far by the packet.
- **Preferred List:** PL which, as already mentioned, is list of nodes determined by the sender node as eligible to forward the Route Request packet.
- **Time to Live (TTL) field:** a decreasing value which determines the allowed duration of packet existence in the network. Every time, before forwarding the Route Request packet, the node that receives the Route Request packet updates it's PLT and inserts in the PL field of the packet, its own PL (which contains the first K nodes from its PLT). The PL of the received packet is always replaced by the one of the receiver.

The Route Request packet is always broadcasted to all nodes neighbors. However, not all these nodes are considered to be eligible to forward the packet further. In order to be eligible a receiver node has to be in the PL of the received packet, not to have forwarded the same packet again and the TTL value must be greater than Zero. If the node satisfies all these requirements, it can forward the packet to all its neighbor nodes.

Before broadcasting the packet however, the node checks whether the destination node is in its NT or NTT. If it is, it unicasts the packet directly to the neighbor which can be the destination node or can just have the destination node as a neighbor in its NT. Otherwise, the packet will be broadcasted with new PL and new parameters. As soon as the destination node is detected, the route selection procedure is followed. As it is obvious multiple RouteRequest packets can reach the destination node. The best route is selected based on the shortest path, the least delay path or the most stable path.

After receiving the first Route Request packet, the destination node starts a timer which indicates for how long it will be receiving Route Request packets from the network. After the expiration of the timer, no more Route Request packets are received by the destination node and the route selection phase can start. Finally in order to deal with link breaks, PLBR uses a particular mechanism to achieve optimal route maintenance. PLBR uses a quick route repair mechanism to bypass the broken link using information about the next two hops from NNT.

But how is the preferred list (PL) constructed? Every time the node receives a Route Request packet, it has to determine its own preferred list of eligible nodes and to replace the received with the new one. Two different algorithms have been proposed for the selection of eligible nodes. The first one is called Neighbor Degree-based Preferred Link Algorithm and is based on neighbor nodes' degree information. The second one is called Weight-based Preferred Link algorithm. Here, the preferred list is calculated based on the stability of the links indicated by the weight. Neighbor Degree-based Preferred Link Algorithm is based on neighbor nodes' degree. Degree of a node is the number of its neighbors. The algorithm is executed for one node and takes under consideration parameters that have to do with the degree of its neighbor nodes. After the execution of the algorithm, two lists are created:

- INL
- EXL
  - Include List (INL) contains a set of neighbor nodes which are considered as reachable for forwarding the Route Request packet.
  - EXL is the Exclude List which contains the set of neighbor nodes which are considered as unreachable.

Advantages

- The efficient flooding mechanism -> reduces the routing control overhead and provides better solutions than the other reactive protocols
- A flooding efficient protocol has higher scalability and decreases network collisions.

Disadvantages

- Both PLBR and WBPL are much more computationally complex than the other reactive protocols.

3.1.11. High Availability Seamless Redundancy Protocol (HSR)

Like PRP, the High Availability Seamless Redundancy Protocol (HSR) applies the principle of parallel operation to a ring interconnecting IEDs with two links. HSR relies on ring topology, sending data not over two networks but in both directions of the single ring.

Like PRP, the High Availability Seamless Redundancy Protocol (HSR) applies the principle of parallel operation to a ring interconnecting IEDs with two links. HSR relies on ring topology, sending data not over two networks but in both directions of the single ring.

In other words, if an IED fails, its protection function and the communication redundancy are lost, but the other IEDs continue to communicate, protect, and control the electrical equipment through a single communication network.

Because HSR forwards all messages from all IEDs, the communication bandwidth is proportional to the number of IEDs. Each device needs to send not only its own messages but also pass along all messages coming from all other IEDs. Therefore, HSR is limited to about 16 communicating devices.

- Cost effective redundancy with no single point of failure and zero recovery time
- Fulfill The dependability and real-time requirements of the most demanding applications such as substation automation and motion control
- Protocol-independent, applicable to most industrial Ethernet

- Do not require switches.
- Each frame has an HSR Ethertype, a path indicator, a size field and a sequence number, inserted as an HSR tag in the same way a VLAN tag is inserted.
- the sender inserts the same sequence number in both frames of a pair, and increments the sequence counter by one for each sending from this node.
- the receiver keeps track of the sequence counter for each source MAC address it receives frames from. Frames with the same source and sequence number value coming from different lines are discarded.
- to supervise the network, a node may keep a table of all other nodes in the network from which it receives frames. This allows to detect nodes absence and bus errors at the same time.
- a node recognizes the frame it sent through its source address and sequence number.
- each node increments the sequence number field monotonically for each frame sent.
- A duplicate frame is recognized in a receiver or forwarding node by its: source address
- sequence number in the HSR tag.
- HSR nodes shall never reject a frame that they did not receive before and shall detect nearly all duplicates, but infrequent duplicates do not disturb.

  The duplicate detection algorithm is not specified. Hash tables, queues and tracking of sequence numbers are possible methods.

  PRP (IEC 62439-3) only considered discard of duplicates on a "best effort" basis. HSR has an improved coverage.

  Each node increments the sequence number field monotonically for each frame sent.

  A duplicate frame is recognized in a receiver or forwarding node by its:

- source address
- sequence number in the HSR tag.
- HSR nodes shall never reject a frame that they did not receive before and shall detect nearly all duplicates, but infrequent duplicates do not disturb.

Core-Extraction Distributed AD HOC Routing Algorithm

CEDAR protocol is a distributed routing protocol oriented to quality of service (QoS) routing in MANET, and bandwidth is the QoS parameter of interest in this protocol. However, the storage and processing overhead of CEDAR is fairly high because too many kinds of control packets are exchanged between nodes and too much state information needs to be maintained by core nodes. The routing algorithm depends fully on the link state information known by core nodes. But the link state information may be imprecise, which results in route failures.

Writing specs is very similar to writing OCUnit tests, except that the language is biased towards describing the behavior of objects. Cedar specs also allows the nesting of contexts so that it is easier to understand how the object behaves in different scenarios.

Creating a New Spec File

1. Create a new file in the spec target.
2. Select the "Cedar Spec" template appropriate for the platform on which the tests will run.
3. Enter the name of the class for which the spec is written.

Global State Routing

- Global State Routing (GSR) is similar to DSDV. It takes the idea of link state routing but improves it by avoiding flooding of routing messages.
- "Global State Routing" (GSlt), where nodes exchange vectors of link states among their neighbors during routing information exchange.
- Based on the link state vectors, nodes maintain a global knowledge of the network topology and optimize their routing decisions locally.
- The performance of the algorithm, studied in this paper through a series of simulations, reveals that this scheme provides a better solution than existing approaches in a truly mobile, ad-hoc environment.

- To design a routing scheme that is MAC efficient in ad-hoc wireless radio networks.
- That is, the control packet size should be able to achieve optimized MAC throughput, and the number of control packets should be controllable.
- It is preferable to maintain the knowledge of full network topology as in link state routing, but wish to avoid the inefficient flooding mechanism.
- Therefore, the scheme is developed based on LS, which has the advantage of routing accuracy, and the dissemination method used in DBF is adopted, which has the advantage of no flooding.
- This scheme is called "Global State Routing" (GSR), and more detailed description is given below.
- Global State Routing (GSR) is a uniform, topologyoriented, proactive routing protocol.
- It is a variant of traditional link-state protocols, in which each node sends link-state information to every node in the network each time its connectivity changes. GSR reduces the cost of disseminating link-state information by relying on periodic exchange of sequenced data rather than flooding.
- In GSR, each node periodically broadcasts its entire topology table to its immediate neighbors.
- The topology table includes the node's most recent assessment of its local connectivity and its current link-state information for the whole network topology. Each entry is tagged with a sequence number.
- A destination's link-state entry is replaced only if the received entry has a larger sequence number.
- Based on the complete topology information in the topology table, any shortest-path algorithm can be used to compute a routing table containing the optimal nexthop information for each destination.
- GSR defines a variant of Dijkstra's algorithm for this purpose.
- As stated earlier, proactive routing protocols maintain routes to all destinations, regardless of whether or not these routes are needed.
- In order to maintain correct route information, a node must periodically send control messages.
- Therefore, proactive routing protocols may waste bandwidth since control messages are sent out unnecessarily when there is no data traffic.

- The main advantage of this category of protocols is that hosts can quickly obtain route information and quickly establish a session.
- For example, GSR introduced below is a proactive routing protocol. Global State Routing (GSR) is based on the Link State (LS) routing method.
- In the LS routing method, each node floods the link state information into the whole network (global flooding) once it realises that links change between itself and its neighbours.
- The link state information includes the delay to each of its neighbours. A node will know the whole topology when it obtains all link information.
- LS routing works well in networks with static topologies. When links change quickly, however, frequent global flooding will inevitably lead to huge control overhead.
- Unlike the traditional LS method, GSR does not flood the link state packets. Instead, every node maintains the link state table based on up-to-date LS information received from neighbouring nodes, and periodically exchanges its LS information with its neighbours only (no global flooding).
- Before sending an LS packet, a node assigns the LS packet a unique sequence number to identify the newest LS information.
- LS information is disseminated as the LS packets with larger sequence numbers replace the ones with smaller sequence numbers.
- The convergence time required to detect a link change in GSR is shorter than in the Distributed Bellman-Ford (DBF) protocol.
- The convergence time in GSR is O(D*I) where D is the diameter of the network and I is the link state update interval.
- The convergence time is normally smaller than O(N*I) in DBF, where N is the number of nodes in the networks and I is the update interval.
- Since the global topology is maintained in every node, preventing routing loops is simple and easy.
- The drawbacks of GSR are the large size of the update messages, which consume a considerable amount of bandwidth, and the latency of the LS information propagation, which depends on the LS information update interval time. ``Fisheye" technology can be used to reduce the size of update messages.
- In this case, every node maintains highly accurate network information about the immediate neighbouring nodes, with progressively fewer details about farther nodes.

- In this algorithm, each node maintains a Neighbor list, a Topology table, a Next Hop table and a Distance table.
- Neighbor list of a node contains the list of its neighbors (here all nodes that can be heard by a node are assumed to be its neighbors.).
- For each destination node, the Topology table contains the link state information as reported by the destination and the timestamp of the information.
- For each destination, the Next Hop table contains the next hop to which the packets for this destination must be forwarded.
- The Distance table contains the shortest distance to each destination node.
- The routing messages are generated on a link change as in link state protocols.
- On receiving a routing message, the node updates its Topology table if the sequence number of the message is newer than the sequence number stored in the table.
- After this the node reconstructs its routing table and broadcasts the information to its neighbors.

CONCLUSION

- *Dynamic Source Routing (DSR) is a routing protocol for wireless mesh networks. Destination-Sequenced Distance Vector (DSDV) routing protocol is a pro-active, table-driven routing protocol for MANETs*
- *AODV allows for the construction of routes to specific destinations and does not require that nodes keep these routes when they are not in active communication.*
- *The source tree Adaptive Routing Protocol (STAR) was the first proactive routing protocol, which worked with link state information.*
- *The Optimized Link State Routing Protocol (OLSR) is an IP routing protocol optimized for mobile ad hoc networks, which can also be used on other wireless ad hoc networks.*
- *Fisheye State Routing (FSR) is an implicit hierarchical routing protocol. CEDAR protocol is a distributed routing protocol oriented to quality of service (QoS) routing in MANET, and bandwidth is the QoS parameter of interest in this protocol.*

Chapter IV

OVERVIEW OF ADVANCED ROUTING PROTOCOLS

OBJECTIVE

- *To learn about the fundamentals of Zone Routing Protocol*
- *To know the classifications of advanced Routing protocols*
- *To become familiar with the concepts of LBR, LAR and WLAR protocols*
- *To understand the application of PLBR and FSMA*
- *To interpret the concepts of Power aware protocols*

4.0. ZONE ROUTING PROTOCOL

- **Zone Routing Protocol**, or **ZRP** is a hybrid Wireless Networking routing protocol that uses both proactive and reactive routing protocols when sending information over the network.
- ZRP was designed to speed up delivery and reduce processing overhead by selecting the most efficient type of protocol to use throughout the route.

Principle of ZRP

- Changes in network topology has local effect only

- Intra-Zone Routing Protocol (IARP)
- Could be any link state or distance vector routing protocol.
- Maintained only within a zone.
- The link-state maintenance gives the ability to have QoS.
- If it is too large, then updates will be too much.
- If it is too small, re-active methods will have to be resorted to.

ZRP Works

- If a packet's destination is in the same zone as the origin, the proactive protocol using an already stored routing table is used to deliver the packet immediately.
- If the route extends outside the packet's originating zone, a reactive protocol takes over to check each successive zone in the route to see whether the destination is inside that zone.
- This reduces the processing overhead for those routes. Once a zone is confirmed as containing the destination node, the proactive protocol, or stored route-listing table, is used to deliver the packet.
- In this way packets with destinations within the same zone as the originating zone are delivered immediately using a stored routing table.
- Packets delivered to nodes outside the sending zone avoid the overhead of checking routing tables along the way by using the reactive protocol to check whether each zone encountered contains the destination node.
- Thus ZRP reduces the control overhead for longer routes that would be necessary if using proactive routing protocols throughout the entire route, while eliminating the delays for routing within a zone that would be caused by the route-discovery processes of reactive routing protocols.
- What is called the Intra-zone Routing Protocol (IARP), or a proactive routing protocol, is used inside routing zones.
- What is called the Inter-zone Routing Protocol (IERP), or are active routing protocol, is used between routing zones.
- IARP uses a routing table. Since this table is already stored, this is considered a proactive protocol. IERP uses a reactive protocol.

- Any route to a destination that is within the same local zone is quickly established from the source's proactively cached routing table by IARP.
- Therefore, if the source and destination of a packet are in the same zone, the packet can be delivered immediately
- Most existing proactive routing algorithms can be used as the IARP for ZRP.
- In ZRP a zone is defined around each node, called the node's k-neighborhood, which consists of all nodes within k hops of the node.
- Border nodes are nodes which are exactly khops away from a source node.
- For routes beyond the local zone, route discovery happens reactively. The source node sends a route request to the border nodes of its zone, containing its own address, the destination address and a unique sequence number.
- Each border node checks its local zone for the destination.
- If the destination is not a member of this local zone, the border node adds its own address to the route request packet and forwards the packet to its own border nodes.
- If the destination is a member of the local zone, it sends a route reply on the reverse path back to the source.
- The source node uses the path saved in the route reply packet to send data packets to the destination.
- Routing protocols for mobile ad-hoc networks have to face the challenge of frequently changing topology, low transmission power and asymmetric links.
- Both proactive and reactive routing protocols prove to be inefficient under these circumstances.
- The Zone Routing Protocol (ZRP) combines the advantages of the proactive and reactive approaches by maintaining an up-to-date topological map of a zone centered on each node.
- Within the zone, routes are immediately available. For destinations outside the zone, ZRP employs a route discovery procedure, which can benefit from the local routing information of the zones.
- In the Zone Routing Protocol, problem of routing in ad-hoc networks and the motivation of ZRP are discussed.
- The architecture of ZRP, which consists of three sub-protocols is described with the routing process illustrated with an example.

- Further, the query control mechanisms are described, which are used to reduce the traffic amount in the route discovery procedure.
- ZRP does not define the actual implementation of the protocol components. Therefore, the guidelines for implementation are presented and example implementations are provided in the draft specifications.
- The problem of routing in networks with unidirectional links, and the proposal for a solution to it are discussed.
- The overhead of the routing protocol is important in the power and bandwidth limited ad-hoc networks.
- The factors influencing the traffic amount based on measurements performed in a number of places are discussed.
- The significant issue of choosing an optimal zone radius, and two algorithms for automatic selection of the radius are also described.

Motivation

- As seen, proactive routing uses excess bandwidth to maintain routing information, while reactive routing involves long route request delays.
- Reactive routing also inefficiently floods the entire network for route determination.
- The Zone Routing Protocol (ZRP) aims to address the problems by combining the best properties of both approaches.
- ZRP can be classed as a hybrid reactive/proactive routing protocol.
- In an ad-hoc network, it can be assumed that the largest part of the traffic is directed to nearby nodes.
- Therefore, ZRP reduces the proactive scope to a zone centered on each node. In a limited zone, the maintenance of routing information is easier.
- Further, the amount of routing information that is never used is minimized. Still, nodes farther away can be reached with reactive routing.
- Since all nodes proactively store local routing information, route requests can be more efficiently performed without querying all the network nodes.
- Despite the use of zones, ZRP has a flat view over the network. In this way, the organizational overhead related to hierarchical protocols can be avoided.

- Hierarchical routing protocols depend on the strategic assignment of gateways or landmarks, so that every node can access all levels, especially the top level.
- Nodes belonging to different subnets must send their communication to a subnet that is common to both nodes.
- This may congest parts of the network. ZRP can be categorized as a flat protocol because the zones overlap.
- Hence, optimal routes can be detected and network congestion can be reduced.
- Further, the behavior of ZRP is adaptive.
- The behavior depends on the current configuration of the network and the behavior of the users.
- Invented by Zygmunt Haas of Cornell University, it combines both reactive and proactive schemes.
- Finds loop free routes to the destination.
- Flat structure (as opposed to hierarchical) is used because Hierarchical schemes can lead to congestion localization.
- Link-State routing is possible – can enable QoS.

Advantages

- Provides some notion of scalability.
- The absence of hierarchies eliminates definitive points of congestion.

4.1. Zone-Based Hierarchical Link State Routing Protocol

- In ZHLS, all network nodes construct two routing tables, an intra-zone routing table and an inter-zone routing table, by flooding Node LSPs within the zone and Zone LSPs throughout the network.
- However, this incurs a large communication overhead in the network. In ZHLS-GF, the flooding scheme floods Zone LSPs only to the gateway nodes of zones and thus reduces the communication overhead significantly.

- Furthermore in ZHLS-GF, only the gateway nodes store Zone LSPs and construct inter-zone routing tables. Therefore the total storage capacity required in the network is less than ZHLS.
- Zone-based Hierarchical Link State routing protocol, formally known as the ZHLS, due to different approach to the routing protocol, is a hierarchical protocol, where the network is divided into non-overlapping zones.
- In addition, mobile nodes are assumed to know their physical locations with assistance from a locating system like GPS.
- Each node only knows the node connectivity within its zone and the zone connectivity of the whole network.
- The zone level topological information is distributed to all nodes. All network nodes in ZHLS construct two routing tables, an intrazone routing table and an interzone routing table.
- ZHLS uses a hierarchical address scheme which contains zone ID and node ID. It is assumed that a virtual link connects two zones if there exists at least one physical link between the zones.
- A two-level network topology structure is defined in ZHLS-the node level topology and the zone level topology.
- In the same way, there are two kinds of link state updates- the node level LSP (Link State Packet) and the zone level LSP.
- A node periodically broadcast its node level LSP to all other nodes in the same zone. In ZHLS, gateway nodes broadcast the zone LSP throughout the network whenever a virtual link is broken or created.
- Consequently, every node knows the current zone level topology of the network. Before sending packets, a source firstly checks its intra-zone routing table.
- If the destination is in the same zone as the source, the routing information is already there.
- Otherwise, the source sends a location request to all other zones through gateway nodes. After a gateway node of the zone, in which the destination node resides, receives the location request, it replies with a location response containing the zone ID of the destination.
- The zone ID and the node ID of the destination node will be specified in the header of the data packets originated from the source.
- During the packet forwarding procedure, intermediate nodes except nodes in the destination zone will use inter-zone routing table, and when the packet arrives the destination zone, an intra-zone routing table will be used.

Advantages

- No overlapping zones
- The zone-level topology information is distributed to all nodes Reduces the traffic and avoids single point of failure Disadvantages: Additional traffic produced by the creation and maintenance of the zone-level topology.

ZHLS

- Info acquirement- Hybrid
- Maintenance loop free- Non-overlapping zone using geographical information
- Periodic messages requirement- Yes
- In ZHLS, mobile nodes are assumed to know their physical locations with assistance from a locating system like GPS.
- The network is divided into non-overlapping zones based on geographical information. ZHLS uses a hierarchical addressing scheme that contains zone ID and node ID.
- A node determines its zone ID according to its location and the pre-defined zone map is well known to all nodes in the network.
- It is assumed that a virtual link connects two zones if there exists at least one physical link between the zones.
- A two-level network topology structure is defined in ZHLS, the node level topology and the zone level topology.
- Respectively, there are two kinds of link state updates, the node level LSP (Link State Packet) and the zone level LSP.
- A node level LSP contains the node IDs of its neighbours in the same zone and the zone IDs of all other zones.
- A node periodically broadcasts its node level LSP to all other nodes in the same zone.
- Therefore, through periodic node level LSP exchanges, all nodes in a zone keep identical node level link state information.
- In ZHLS, gateway nodes broadcast the zone LSP throughout the network whenever a virtual link is broken or created.

- Consequently, every node knows the current zone level topology of the network. Before sending packets, a source firstly checks its intrazone routing table.
- If the destination is in the same zone as the source, the routing information is already there. Otherwise, the source sends a location request to all other zones through gateway nodes.
- After a gateway node of the zone, in which the destination node resides, receives the location request, it replies with a location response containing the zone ID of the destination.
- The zone ID and the node ID of the destination node will be specified in the header of the data packets originated from the source.
- During the packet forwarding procedure, intermediate nodes except nodes in the destination zone will use inter-zone routing table, and when the packet arrives the destination zone, an intra-zone routing table will be used.
- ZHLS assumes that each node has a location system such as GPS and the geographical information is well known, and the network is geographically divided into non-overlapping zones.
- The performance of a zone based routing protocol is tightly related to the dynamics and size of the network and parameters for zone construction.
- However, because zones heavily overlap, ZRP in general will incur more overhead than ZHLS and HARP
- All three zone-based routing protocols presented in this subsection use proactive routing for intra-zone communication and reactive routing for inter-zone packet forwarding.
- Performance of a zone based routing protocol is decided by the performance of respective proactive and reactive routing protocols chosen and how they cooperate each other.
- A new global positioning system (GPS)-based routing protocol for ad hoc networks, called zone-based hierarchical link state (ZHLS) routing protocol, is proposed.
- In this protocol, the network is divided into non overlapping zones. Each node only knows the node connectivity within its zone and the zone connectivity of the whole network.
- The link state routing is performed on two levels: local node and global zone levels. Unlike other hierarchical protocols, there is no cluster head in this protocol. The zone level topological information is distributed to all nodes.

- This "peer-to-peer" manner mitigates traffic bottleneck, avoids single point of failure, and simplifies mobility management.
- Since only zone ID and node ID of a destination are needed for routing, the route from a source to a destination is adaptable to changing topology.
- The zone ID of the destination is found by sending one location request to every zone. Simulation results show that our location search scheme generates less overhead than the schemes based on flooding.
- The results also confirm that the communication overhead for creating and maintaining the topology in the proposed protocol is smaller than that in the flat LSR protocol.
- This new routing protocol provides a flexible, efficient, and effective approach to accommodate the changing topology in a wireless network environment.
- Similar to ZRP, ZHLS is a hybrid reactive/proactive scheme. It is proactive if the destination is within the same zone of the source.
- Otherwise, it is reactive because a location search is needed to find the zone ID of the destination.
- However, unlike ZRP, ZHLS requires GPS and maintains a high level hierarchy for interzone routing.
- Location search is performed by unicasting one location request to each zone. Routing is done by specifying the zone ID and the node ID of the destination, instead of specifying an ordered list of all the intermediate nodes between the source and the destination.
- Intermediate link breakage may not cause any subsequent location search. Since the network consists of nonoverlapping zones in ZHLS, frequency reuse is readily deployable in ZHLS.
- The main idea is to use the hierarchical routing approach in a peer-to-peer way for large mobile wireless networks.
- Scarce wireless resource is an important factor in designing a routing protocol for mobile ad hoc networks.
- Simulation data has shown that, for a network of size, the amount of communication overhead in the proposed ZHLS is of the order.
- Node mobility is another important design consideration. In ZHLS, node mobility does not usually have a global effect, and not many globally propagated control messages are generated.
- On the other hand, in a flat protocol, node mobility has a global effect and creates a significant amount of globally propagated control messages.

- Simulation results have asserted that the communication overhead induced by mobility is much smaller in ZHLS than that in LSR.
- Also, simulation results have confirmed that the chance of a virtual link breakage in ZHLS is smaller than that of a physical link breakage in a flat protocol.
- So, the zone level topology in ZHLS is relatively stable.
- Unlike other hierarchical routing protocols our peer-to-peer hierarchical protocol does not designate any node as cluster head.
- As a result, a single point of failure and traffic bottleneck can be avoided. Mobility management in ZHLS is simple as all nodes play the same role in the network.
- Owing to the bandwidth constraint, most ad hoc routing protocols are reactive; each node only maintains the routing information of active destinations.
- Therefore, path search is necessary for all reactive schemes. In reactive schemes with a flat architecture, flooding is the only way to search for a route to a destination.
- On the other hand, in ZHLS that has a hierarchical architecture, broadcasting on the zone level topology is used for location searches.
- Simulation results demonstrate that broadcasting on the zone level topology in ZHLS generates a smaller amount of communication overhead than flooding in a flat protocol. Moreover, in the flat and reactive protocols, path containing the nodes between a source and a destination is needed for routing.
- Any intermediate link breakage will invalidate the path and render subsequent search. On the contrary, in ZHLS, only the zone ID and the node ID of a destination are needed for routing.
- The actual routing path is adaptable to the changing topology, and a subsequent search is not required as long as the destination does not hand off to another zone. Handoff management has been included in ZHLS to alleviate the handoff effect.
- In ZHLS, the network is divided into zones as in cellular networks. Frequency reuse commonly used in cellular networks is readily deployable in our protocol to lessen the bandwidth constraint.
- Furthermore, handoff concept is borrowed from the cellular networks to design handoff management.

4.2. Route-Lifetime Assessment–Based Routing (RABR)

- The protocol requires that each node maintains a record of the affinities of its neighbors as proposed. Various aspects such as route discovery, route selection and route maintenance are presented in detail.

Route Discovery

- RABR is an On-demand protocol and the route discovery process similar to DSR.
- When a source has a packet to transmit to a destination for which there is no entry in the routing table, a route request is initiated.
- The route establishment is done by broadcasting mute request packets that contain a sequence number, source id and the destination id.
- An intermediate host listening to the request broadcasts the request further, after appending the affinity of the link from which it received the packet.
- Any further requests with the same are not entertained by this node.
- This process continues until the request reaches the destination. After receiving the first request, the destination waits fora fixed time interval for more mule request packets (carrying the source to destination routes) to arrive.
- The methodology of selection of an optimal route out of the candidate routes for non-TCP and TCP traffic respectively.
- A route cache has not been included in current implementation, i.e., an intermediate node does not dispatch a route reply to the source, even if it maintains a route entry to that destination.

4.3. LBR (Load Balanced Routing Protocols)

Load balancing deals with improving the performance of the system by transferring the jobs from over headed nodes to under loaded or idle nodes. In ad hoc networks, only Associatively-Based Routing (ABR) considers the loads the metric. ABR, however, uses the routing load as the secondary metric.

Furthermore, the load is measured in the number of routes a node is a part of, and hence the protocol does not account for various traffic loads of each date session.

4.4. Dynamic Load Aware Routing (DLAR)

DLAR considers the load of intermediate nodes as the main route selection metric and monitors the congestion status of active routes to reconstruct the path when nodes of the route have their interface queue overloaded.

DLAR builds routes on-demand. When a route is required but no information to the destination is known, the source floods the ROUTE REQUEST packet to discover a route. When nodes other than the destination receive a no duplicate ROUTE REQUEST, they build a route entry for the <source, destination> pair and record the previous hop to that entry (thus, backward learning).

Nodes then attach their load information (the number of packets buffered in their interface) and broadcast the ROUTE REQUEST packet. After receiving the first ROUTE REQUEST packet, the destination waits for an appropriate amount of time to learn all possible routes. In order to learn all the routes and their quality, the destination node accepts duplicate ROUTE REQUESTS received from different previous nodes. The destination then chooses the least loaded route and sends a ROUTE REPLY packet back to the source via the selected route.

A node can detect a link break by receiving a link layer feedback signal from the MAC protocol, not receiving passive acknowledgment, or not receiving hello packets for a certain period of time. When a route is disconnected, the immediate upstream node of the broken link sends a ROUTE ERROR message to the source of the route to notify the route invalidation. Nodes along the path to the source remove the route entry upon receiving this message and relay it to the source. The source reconstructs a route by flooding a ROUTE REQUEST when informed of a route disconnection.

DLAR introduce three algorithms in selecting the least loaded route. DLAR scheme 1 simply adds the routing load of each intermediate node and selects the route with the least sum. If there is a tie, the destination selects the route with the shortest hop distance. When there are still multiple routes that have the least load and hop distance, the path that is taken by the packet which arrived at the destination the earliest between them is chosen. DLAR scheme 2 is similar to scheme 1. However, instead of using the sum of number of packets queued at each intermediate node's interface as in scheme 1, scheme 2 uses the average number of packets buffered at each intermediate node along the path. DLAR scheme 3 considers the number of congested intermediate nodes as the route selection metric. Basically, it chooses the route with the least number of intermediate nodes that have their load exceeding the threshold value T.

DLAR does not allow intermediate nodes to reply from cache. DLAR periodically monitors the congestion status of active data sessions and dynamically reconfigures the routes that are being congested. Using the least-loaded routes helps balance the load of the network nodes and utilize the network resources efficiently.

4.5. Load-Aware Destination-Controlled Routing for Manet (LBAR)

LBAR defines a new metric for routing known as the degree of nodal activity to represent the load on a mobile node. In LBAR routing information on all paths from source to destination are forwarded through setup messages to the destination. Setup messages include nodal activity information of all nodes on the traversed path.

After collecting information on all possible paths, the destination then makes a selection of the path with the best-cost value and sends an acknowledgement to the source node. LBAR also provides an alternate path maintenance technique to patch up broken links by detouring traffic to the destination.

Load-Balanced Ad hoc Routing (LBAR) is an on-demand routing protocol intended for delay-sensitive applications where users are most concerned with packet transmission delay. Hence, LBAR focuses on how to find a path, which would reflect least traffic, load so that data packets can be routed with least delay. The algorithm has four components: Route Discovery, Path Maintenance, Local Connectivity Management, Cost Function Computation. First Route recovery, the route discovery process is initiated whenever a source node needs to communicate with another node for which it does not have a known route. The process is divided into two stages: forward and backward. The forward stage starts at the source node by broadcasting setup messages to its neighbors.

A setup message carries the cost seen from the source to the current node. A node that receives a setup message will forward it, in the same manner, to its neighbors after updating the cost based on its nodal activity value. The backward stage begins with an ACK message forwarded backward towards the source node along the selected path, which is called the active path. If a link on the selected path breaks, the ACK message is discarded and an error message is sent backward along the path fragment to the destination. The destination node will then choose another path.

Second Path maintenance, if the source node, an intermediate node on the active path or the destination node moves out of the communication range, an alternate path must be found. If the source node moves away from the active path, In that case, the source has to reinitiate the route discovery procedure to establish a new route to the destination. When either the destination node or some intermediate node moves outside the active path, path maintenance will be initiated to correct the broken path. Once the next hop becomes unreachable, the node upstream of the broken hop propagates an error message to the destination node. Upon receiving notification of a broken link, the destination node picks up an alternative best-cost partial route passing through the node propagating the error message and then sends an ACK message to the initiator of the error message. If the destination has no alternative path passing through the node sending the error message, the destination picks up another route and sends an ACK message to the source. The source will use this new route to send data packets if it still has data to send. By then, a new active path is defined. In the worst case, where the destination has no alternate paths, it propagates an error message to the source and lets it restart route discovery.

Third Local connectivity management, whenever a node receives a broadcast from a neighbor, it updates its local connectivity information in its Neighborhood table to ensure that it includes this neighbor. Source broadcasts a hello message to its neighbors, containing its identity and activity. Neighbors that receive this packet update their local connectivity information in their Neighborhood tables. Receiving a broadcast or a hello from a new neighbor, or failing to receive consecutive hello messages from a node previously in the neighborhood, is an indication that the local connectivity has changed. If hello messages are not received from the next hop along an active path, the upstream active neighbors using that next hop send notification of link failure and the path maintenance protocol is invoked. The cost function is used to find a path with the least traffic so that data packets can be transmitted to the destination as fast as possible which achieves the goal of balancing loads over the network. In this protocol, Active path is a path from a source to a destination, which is followed by packets along this selected route.

Active node is considered active if it originates or relays data packets or is a destination. Inactive node is considered inactive if it is note along an active path. Activity is the number of active paths through a node is defined as a metric measuring the activity of the node. Cost is the minimum traffic load plus interference is proposed as the metric for best cost. Unlike wired networks, packet delay is not caused only from traffic load at the current node, but also by traffic load at neighboring nodes.

We call this traffic interference. In the contest of traffic interference, the best cost route is regarded as a path, which encounters the minimum traffic load in transmission and minimum interference by neighboring nodes. To assess best cost, the term node activity is used as an indirect means to reflect traffic load at the node. Such activity information can be gained at the network layer, independent of the MAC layer. Traffic interference is defined as the sum of neighboring activity of the current node. During the routing stage, nodal activity and traffic interference are calculated at every intermediate node along path from source to destination. When the destination received routing information, it chooses a path, which has minimum cost.

4.6. Load Aware Routing in AD HOC Networks (LARA)

Load Aware Routing in Ad hoc (LARA) networks protocol for efficient data transmission in mobile ad hoc networks. A new metric for routing called traffic density to represent the degree of contention at the medium access control layer. During the route setup, this metric is used to select the route with the minimum traffic load. LARA protocol requires that each node maintain a record of the latest traffic queue estimations at each of its neighbors in a table called the neighborhood table. This table is used to keep the load information of local neighbors at each node.

This information is collected through two types of broadcasts. The first type of broadcast occurs when a node attempts to discover route to a destination node. This type of broadcast is called route request. The second type of broadcasting is the hello packet broadcasting. In the event that a node has not sent any messages to any of its neighbors within a predefined timeout period, called the hello interval, it broadcasts a hello message to its neighbors.

A hello packet contains the sender node's identity and its traffic queue status. Neighbors that receive this packet update the corresponding neighbor's load information in their neighborhood tables.

If a node does not receive a data or a hello message from some of its neighbors for a predefined time, it assumes that these nodes have moved out of the radio range of this node and it changes its neighborhood table accordingly. Receiving a message from a new node is also an indication of the change of neighbor information and is handled appropriately. Traffic queue. The traffic queue of a node is defined as the average value of the interface queue length measured over a period of time.

4.7. Delay-Based Load-Aware on-Demand Routing (D-LOAR)

Delay-based Load-Aware On-demand Routing (D-LAOR) protocol, which determines the optimal path based on the estimated total path delay and the hop count. D-LAOR scheme that utilizes both the estimated total path delay and the hop count as the route selection criterion. D-LAOR also has a mechanism in new route selection to avoid a congested node by selectively dropping the Route Request (RREQ) packets. Simulation results show that our proposed D-LAOR scheme increases packet delivery fraction and decreases end-to-end delay by more than 10% in a moderate network scenario when compared with the original AODV and other LAOR protocols.

D-LAOR can route around a congested node and thus can reduce the control overhead. This is achieved by dropping the RREQ packets at congested nodes, which prevents the congested node from becoming an intermediate node of a path. D-LAOR determines the congested node by comparing the estimated total node delay and the number of packets being queued in the interface queue of two serial nodes in a RREQ packet-forwarding path. DLAOR drops a RREQ packet only when the following two conditions are satisfied simultaneously,

1. The estimated total node delay of a node A is greater than that of previous node B.
2. The number of packets being queued at the interface queue of a node A is more than 80% of its buffer size.

4.8. Weighted Load Aware Routing (WLAR)

However, these routing protocols reflect neither burst traffic nor transient congestion. To work out this problem, Weighted Load Aware Routing (WLAR) protocol is proposed. This protocol selects the route based on the information from the neighbor nodes which are on the route to the destination. In WLAR, a new term traffic load is defined as the product of average queue size of the interface at the node and the number of sharing nodes which are declared to influence the transmission of their neighbors.(WLAR) protocol adopts basic AODV procedure and packet format.

In WLAR, each node has to measure its average number of packets queued in its interface, and then check whether it is a sharing node to its neighbor or not. If it is a sharing node itself, it has to let its neighbors know it. After each node gets its own average packet queue size and the number of its sharing nodes, it has to calculate its own total traffic load. Now when a source node initiates a route discovery procedure by flooding RREQ messages, each node receiving an RREQ will rebroadcast it based on its own total traffic load so that the flooded RREQ's which traverse the heavily loaded routes are dropped on the way or at the destination node. Destination node will select the best route and replies RREP. Average number of packets queued in interface is calculated by Exponentially Weighted Moving Average (EWMA).

The reason to use average number of packets queued in interface is to avoid the influence of transient congestion of router. Sharing node is defined as nodes whose average queue size is greater than or equal to some predetermined threshold value. Sharing node is expected to give some transmission influence to its neighbors. If its average queue size is not greater than a threshold value, it is assumed that its effect is negligible. Total traffic load in node is defined as its own traffic load plus the product of its own traffic load and the number of sharing nodes. Path load is defined as sum of total traffic loads of the nodes which include source node and all intermediate nodes on the route, except the destination node.

4.9. Simple Load-Balancing AD HOC Routing (SLAR)

Simple Load-Balancing Ad hoc Routing (SLAR) protocol is based on the autonomy of each node. Although it may not provide the network-wide optimized solution but it may reduce the overhead incurred by load balancing and prevent from severe battery power consumption caused by forwarding packets. In SLAR, each node determines whether it is under heavy forwarding load condition, and in that case it gives up forwarding packets and lets some other nodes take over the role. In MANETs, since nodes have limited resources, the message overhead for load balancing is more critical than that of the wired network, i.e., in the ad hoc network, the network-wide optimized load balancing approach of the wired network may be inappropriate. SLAR is designed not as an entirely new routing protocol but as an enhancement of any existing ad hoc routing protocols like AODV, DSR etc.

4.10. PAR (Power Aware Routing Protocol)

The main objective of power aware routing protocols is to minimize the power consumption and maximize the network lifetime. The network lifetime is defined up to the moment when a node runs out of its own battery power for the first time. If a node stops its operation, it can result in network partitioning and interrupt communication.

Power Aware Routing (PAR) protocol gives maximum network lifetime and minimum power consumption by selecting less congested and more stable route, during the source to destination route establishment process, to transfer real-time and non real-time traffic, hence providing energy efficient routes. PAR focuses on 3 parameters: Accumulated energy of a path, Status of battery lifetime and Type of data to be transferred. At the moment of route selection, PAR focuses on its core metrics like traffic level on the path, battery status of the path, and type of request from user side. With these factors in consideration, PAR always selects less congested and more stable routes for data delivery and can provide different routes for different type of data transfer and ultimately increases the network lifetime. Simulation results shows that PAR is similar such as DSR and AODV protocols, with respects to different energy -related performance metrics even in high mobility scenarios. Although, PAR can somewhat incur increased latency during data transfer, it discover routed that can last for a long time and encounter significant power saving.

4.11. Optimized Link State Routing Protocol

- The **Optimized Link State Routing Protocol (OLSR)** is an IP routing protocol optimized for mobile ad hoc networks, which can also be used on other wireless ad hoc networks.
- OLSR is a proactive link-state routing protocol, which uses hello and topology control (TC) messages to discover and then disseminate link state information throughout the mobile ad hoc network.
- Individual nodes use this topology information to compute next hop destinations for all nodes in the network using shortest hop forwarding paths.
- Link-state routing protocols such as Open Shortest Path First (OSPF) and IS-IS elect a designated router on every link to perform flooding of topology information.
- In wireless ad hoc networks, there is different notion of a link, packets can and do go out the same interface; hence, a different approach is needed in order to optimize the flooding process.
- Using Hello messages the OLSR protocol at each node discovers 2-hop neighbor information and performs a distributed election of a set of multipoint relays (MPRs). Nodes select MPRs such that there exists a path to each of its 2-hop neighbors via a node selected as an MPR.
- These MPR nodes then source and forward TC messages that contain the MPR selectors.
- This functioning of MPRs makes OLSR unique from other link state routing protocols in a few different ways: The forwarding path for TC messages is not shared among all nodes but varies depending on the source, only a subset of nodes source link state information, not all links of a node are advertised but only those that represent MPR selections.
- Since link-state routing requires the topology database to be synchronized across the network, OSPF and IS-IS perform topology flooding using a reliable algorithm.
- Such an algorithm is very difficult to design for ad hoc wireless networks, so OLSR doesn't bother with reliability.
- It simply floods topology data often enough to make sure that the database does not remain unsynchronized for extended periods of time.

- Default and network routes can be injected into the system by HNA messages allowing for connection to the internet or other networks within the OLSR MANET cloud.
- Network routes are something reactive protocols do not currently execute well.
- Timeout values and validity information is contained within the messages conveying information allowing for differing timer values to be used at differing nodes.
- The original definition of OLSR does not include any provisions for sensing of link quality; it simply assumes that a link is up if a number of hello packets have been received recently.
- This assumes that links are bi-modal (either working or failed), which is not necessarily the case on wireless networks, where links often exhibit intermediate rates of packet loss. Implementations such as the open source OLSRd (commonly used on Linux-based mesh routers) have been extended (as of v. 0.4.8) with link quality sensing.
- Being a proactive protocol, OLSR uses power and network resources in order to propagate data about possibly unused routes.
- While this is not a problem for wired access points, and laptops, it makes OLSR unsuitable for sensor networks that try to sleep most of the time.
- For small scale wired access points with low CPU power, the open source OLSRd project showed that large scale mesh networks can run with OLSRd on thousands of nodes with very little CPU power on 200 MHz embedded devices.
- Being a link-state protocol, OLSR requires a reasonably large amount of bandwidth and CPU power to compute optimal paths in the network.
- In the typical networks where OLSR is used (which rarely exceed a few hundreds of nodes), this does not appear to be a problem.
- By only using MPRs to flood topology information, OLSR removes some of the redundancy of the flooding process, which may be a problem in networks with moderate to large packet loss rates – however the MPR mechanism is self-pruning (which means that in case of packet losses, some nodes that would not have retransmitted a packet, may do so).
- OLSR makes use of "Hello" messages to find its one hop neighbors and its two hop neighbors through their responses.

- The sender can then select its multipoint relays (MPR) based on the one hop node that offers the best routes to the two hop nodes.
- Each node has also an MPR selector set, which enumerates nodes that have selected it as an MPR node.
- OLSR uses topology control (TC) messages along with MPR forwarding to disseminate neighbor information throughout the network.
- Host and network association (HNA) messages are used by OLSR to disseminate network route advertisements in the same way TC messages advertise host routes.
- The problem of routing in ad hoc wireless networks is actively being researched, and OLSR is but one of several proposed solutions.
- To many, it is not clear whether a whole new protocol is needed, or whether OSPF could be extended with support for wireless interfaces.
- In bandwidth- and power-starved environments, it is interesting to keep the network silent when there is no traffic to be routed.
- Reactive routing protocols do not maintain routes, but build them on demand.
- As link-state protocols require database synchronisation, such protocols typically use the distance vector approach, as in AODV and DSDV, or more ad hoc approaches that do not necessarily build optimal paths, such as Dynamic Source Routing.
- The information in this section concerning the Optimized Link State Protocol is taken from its RFC 356.
- Optimized Link State Protocol (OLSR) is a proactive routing protocol, so the routes are always immediately available when needed.
- OLSR is an optimization version of a pure link state protocol. So the topological changes cause the flooding of the topological information to all available hosts in the network.
- To reduce the possible overhead in the network protocol uses Multipoint Relays (MPR). The idea of MPR is to reduce flooding of broadcasts by reducing the same broadcast in some regions in the network, more details about MPR can be found later in this chapter. Another reduce is to provide the shortest path.
- Hello messages are used for finding the information about the link status and the host's neighbours.
- With the Hello message the Multipoint Relay (MPR) Selector set is constructed which describes which neighbours has chosen this host to

act as MPR and from this information the host can calculate its own set of the MPRs.

- The Hello messages are sent only one hop away but the TC messages are broadcasted throughout the entire network. TC messages are used for broadcasting information about own advertised neighbours which includes at least the MPR Selector list. The TC messages are broadcasted periodically and only the MPR hosts can forward the TC messages.
- There is also Multiple Interface Declaration (MID) messages which are used for informing other host that the announcing host can have multiple OLSR interface addresses.
- The MID message is broadcasted throughout the entire network only by MPRs. There is also a "Host and Network Association" (HNA) message which provides the external routing information by giving the possibility for routing to the external addresses.
- The HNA message provides information about the network- and the netmask addresses, so that OLSR host can consider that the announcing host can act as a gateway to the announcing set of addresses.
- The HNA is considered as a generalized version of the TC message with only difference that the TC message can inform about route cancelling while HNA message information is removed only after expiration time.
- The MID and HNA messages are not explained in more details in this chapter, the further information concerning these messages.

Advantages

- Being a proactive protocol, routes to all destinations within the network are known and maintained before use.
- Having the routes available within the standard routing table can be useful for some systems and network applications as there is no route discovery delay associated with finding a new route.
- The routing overhead generated, while generally greater than that of a reactive protocol, does not increase with the number of routes being created.
- OLSR protocol is well suited for the application which does not allow the long delays in the transmission of the data packets.

- OLSR has also extensions to allow for hosts to have multiple OLSR interface addresses and provide the external routing information giving the possibility for routing to the external addresses
- OLSR is also a flat routing protocol, it does not need central administrative system to handle its routing process.
- The proactive characteristic of the protocol provides that the protocol has all the routing information to all participated hosts in the network.

Disadvantages

- OLSR needs more time re-discovering a broken link.
- Wider delay distribution
- OLSR requires more processing power when discovering an alternate route.

4.12. Location-Aided Routing (LAR)

- Main Idea
  - Using location information to reduce the number of nodes to whom route request is propagated.
  - Location-aided route discovery based on "limited" flooding
- Location-Aided Routing (LAR) protocols limit the search for a new route to a smaller "request zone" of the ad hoc network.
- This results in a significant reduction in the number of routing messages

Route Discovery Using Flooding

- The route discovery algorithm using flooding is described next. When a node S needs to find a route to node D, node S broadcasts a route request message to all its neighbors – hereafter, node S will be referred to as the sender and node D as the destination.
- A node, say X, on receiving a route request message, compares the desired destination with its own identifier. If there is a match, it means that the request is for a route to itself (i.e., node X). Otherwise, node

X broadcasts the request to its neighbors – to avoid redundant transmissions of route requests, a node X only broadcasts a particular route request once (repeated reception of a route request is detected using sequence numbers).

- In this figure, node S needs to determine a route to node D. Therefore, node S broadcasts a route request to its neighbors. When nodes B and C receive the route request, they forward it to all their neighbors.
- When node X receives the route request from B, it forwards the request to its neighbors. However, when node X receives the same route request from C, node X simply discards the route request.
- Route discovery is initiated either when the sender S detects that a previously determined route to node D is broken, or if S does not know a route to the destination. In our implementation, we assume that node S can know that the route is broken only if it attempts to use the route.
- When node S sends a data packet along a particular route, a node along that path returns a route error message, if the next hop on the route is broken.
- When node S receives the route error message, it initiates route discovery for destination D. When using the above algorithm, observe that the route request would reach every node that is reachable from node S (potentially, all nodes in the ad hoc network).
- Using location information, reduce the number of nodes to whom route request is propagated.
- Dynamic source routing (DSR) and ad hoc on-demand distance vector routing (AODV) protocols proposed previously are both based on variations of flooding.
- DSR and AODV also use some optimizations - several of these optimizations as well as other optimizations suggested in this paper can be used in conjunction with the proposed algorithms.
- However, for simplicity, Location-Aided Routing (LAR), as it makes use of location information to reduce routing overhead.
- Location information used in the LAR protocol may be provided by the Global Positioning System (GPS).
- With the availability of GPS, it is possible for a mobile host to know its physical location. In reality, position information provided by GPS includes some amount of error, which is the difference between GPS-calculated coordinates and the real coordinates.

- For instance, NAVSTAR Global Positioning System has positional accuracy of about 50-100 meters and Differential GPS offers accuracies of a few meters.
- However, the ideas suggested here can also be applied when the location is known only approximately – the Performance Evaluation section considers this possibility.

Disadvantages

- Possible improvement with more accurate request zones by using GPS; security concerns must be addressed
- A request zone can be calculated at each intermediate node than from the source node
- In LAR 1 scheme, each intermediate node calculate its own request zone and broadcast to all its neighbors. This may further reduce the overhead and route discovery can be initiated from intermediate node to the destination
- In LAR 1 scheme, determining the rectangle is by considering relative positions of source and destination
- Scalability
- Performance comparison of LAR with widely used routing protocols
- LAR and CGSR
    - LAR and AODV

4.13. MACAW PROTOCOL

Multiple Access with Collision Avoidance for Wireless

- Multiple Access with Collision Avoidance for Wireless (MACAW) is a slotted Medium Access Control (MAC) protocol widely used in ad hoc networks.
- Furthermore, it is the foundation of many other MAC protocols used in Wireless Sensor Networks (WSN). The IEEE 802.11 RTS/CTS mechanism is adopted from this protocol.

- It uses RTS-CTS-DS-DATA-ACK frame sequence for transferring data, sometimes preceded by an RTS-RRTS frame sequence, in view to provide solution to the hidden terminal problem.
- Although protocols based on MACAW, such as S-MAC, use carrier sense in addition to the RTS/CTS mechanism, MACAW does not make use of carrier sense.

MACAW: A Media Access Protocol for Wireless LAN's

- Based on MACA it is a Multiple Access Collision Avoidance protocol.
- Initial attempt to deal with WLAN challenges
- Four key main observations
- The relevant contention is at the receiver not the sender.
- Congestion is location dependent.„
- Learning about the contention level must be a collective enterprise.
- The media access protocol should propagate congestion information explicitly
- A sends out RTS and sets a timer and waits for CTS
    - If A receives CTS before timer goes to zero, it is ok and sends data packet
    - Otherwise, A assumes there is a collision at B
        - Doubles the backoff counter interval
            - Randomly picks up a timer from [1, backoff counter]
        - Sends next RTS after timer go to zero
- B sends out CTS, then sets a timer and waits for data packet
    - If data packet arrives before timer goes to zero, it is ok.
    - Otherwise, B can do other things
- C overhears A's RTS, sets a timer which is long enough to allow A to receive CTS. After the timer goes to zero, C can do other things
- D overhears B's CTS, sets a timer which is long enough to allow B to receive data packet.
- E overhears A's RTS and B's CTS, sets a timer which is long enough to allow B to receive data packet.
- RTS and CTS can also contain info to allow sender A to adjust power to reduce interference
- MACA did not solve hidden & exposed terminal problems
- MACA did not provide specifications about parameters such as

  - What are the RTS, CTS packet sizes.
  - How to decide timers.
  - What is the size of the initial backoff window.
- A lot things need to be done using MACA

Goals, New Ideas, and Main Contributions

- Goals:
  - It is used to extend MACA
- New Idea: Information sharing to achieve fairness
- Main Results:
  - Modified control messages
    - Four-way handshake (reliable, recovery at MAC layer)
    - Five-way handshake (relieve exposed terminal problem)
    - RRTS (unfairness)
  - Modified back-off algorithms
    - Multiplicative increase and linear decrease (MILD)
    - Synchronize back-off counter using piggyback message
  - Multiple stream model (V-MAC)

Three-Way Handshake

- A sends Ready-to-Send (RTS)
- B responds with Clear-to-Send (CTS)
- A sends DATA PACKET
- RTS and CTS announce the duration of the data transfer
- Nodes overhearing RTS keep quiet for some time to allow A to receive CTS
- Nodes overhearing CTS keep quiet for some time to allow B to receive data packet
- A "media access protocol" is a link-layer protocol for allocating access to a shared medium (i.e., a wireless channel) among multiple transmitters.
- MACAW is a media access protocol for mobile pads that communicate wirelessly with base stations; each base station can hear

transmission from pads within a certain range of its location, called a cell.

This is a challenging problem for several reasons:

- Transmitters are mobile, but location information is not communicated explicitly: a transmitter simply moves or a new transmitter comes into the area, and the media access protocol must adapt.
- Connectivity is not transitive. For example, A and B can communicate, as can B and C, but A and C might not be able to; the protocol must be designed so that communications between B and C do not disrupt A, even though A has no direct knowledge of C.
- Due to noise, communication may not be symmetric (A can hear B, but B cannot hear A), and transmitters may interfere with one another (A might be out of range of B, but A's transmissions might still disrupt B's ability to communicate).
- The MACAW paper ignores both asymmetry and interference: they assume a simplified model in which two stations are either in range of one another or they aren't; a transmission is successfully received if only if there is only one active transmitter in range of the receiver; and no two base stations are within range of one another.
- Importantly, contention between transmissions occurs at the receiver of the transmission, not the sender: that is, two nodes can transmit different messages from the same area at once, but two nodes in the same area cannot simultaneously receive different messages.

Their goal is a media access protocol that is both efficient (high utilization of the available media), and fair (each station gets its "fair share").

4.14. Preferred Link Based Routing (PLBR) Protocol

Working Principles

- The preferred link approach is used in SSA.

- The lack of topology information, link characteristics changing continuously, broadcast nature of the radio channel, and nodes mobility are the reason for not suitable to usethe single preferred link model in ad hoc networks.
- Sisodia et al. proposed two algorithms known as Preferred Link Based Routing (PLBR) protocols. In this protocol a node selects a subset of nodes from the list of its neighbor node [1, 24].
- PLBR (Preferred Link Based Routing) can also exploit link metrics including stability.
- Different from the above two protocols, PLBR adopts a neighbor selection technique in forwarding RREQ messages where only preferred neighbors are allowed to forward messages in order to avoid broadcast storm.
- The collected values are used to determine the best route. Our methods are different from the above methods in the following points.
- First, it is aimed at providing not only information dissemination but also communication infrastructure between vehicles or between roadsides and vehicles.
- Obviously, routing is mandatory for this purpose, and this goal cannot be achieved by information dissemination protocols for VANETs since their target applications are delay-tolerant data dissemination.
- Secondly, any existing method does not exploit map information to find the route. Moreover, it is assumed that the locations of a source and a destination are fixed.
- Preferred Link-Based Unified (PLBU) Routing Protocol can be used both for unicast and multicast.
- The protocol forwards RouteRequest (RReq) packets to the nearby nodes based on Link Stability, which is being calculated by the power levels of received periodic beacon signals.
- The integrated use of PLBU Routing and D-MAC protocol results in improvement of Throughput to the Network and Bandwidth Efficiency.
- PLBM a source tree based receiver initiated protocol employs hard state approach for tree maintenance.
- Here each node maintains two tables NNT (Neighbors Neighbors Table) maintains two hop (on both streams) local network topology information and CT (Connect Table) for multicast tree information. Every node updates its NNT entry on receiving beacon packets that are transmitted periodically by every node. When a new member

wants to join the group sends JoinConfirm message to the tree nodes in its NNT. Otherwise sends JoinQuery packet to one of the eligible nodes which are determined using Preferred Link Based Algorithm (PLBA).

- Only the preferred nodes (first K eligible nodes in NNT) can process the received JoinQuery packet and forwards to eligible nodes starting the timer to receive JoinConfirm message from the receiver who wish to join the group.
- Each intermediate node updates its CT (path information both on upstream and downstream) on receiving JoinConfirm packet and marks itself as connected.
- The concept of the preferred link provides better adaptability and flexibility.
- Further, the use of 2 hop local topology information provides efficient multicast routing.
- The preferred list not only relies on neighboring nodes but also on other characteristics such as link delay, bandwidth, and link stability, network load.

Advantages of PLBR

- On demand routing protocols have the problem of broadcast minimized in this protocol.
- Decrease in the number of collisions as well as improvement in the efficiency of the protocol leads to reduction in control overhead.

Disadvantages

- Transmission of beacon packets periodically increases control overhead leading
- More complex for computation of NDPL (Neighbor Degree based preferred Link algorithm) and WBPL (Weight Based Preferred Link algorithm) than other forwarding schemes of Route Request.

Fading Signature Multiple Access (FSMA)

Fading signature multiple access (FSMA) is a promising radio access technology that utilizes the differ-ence in users' fading channels for joint signal detection. However, channels' correlation has direct impact to degrade the capacity of wireless communications considerably. Here, efficient signal design methods are proposed to mitigate the correlation effects and maintain highest achievable channel capacity of correlated FSMA system. Two methods named unequal power signal constellation (UPSC) and rotated signal constellation (RSC) are considered and compared with the conventional identical signal constellation (ISC) technique. Based on the principles of maximizing the minimum Euclidean distance of composite received signals, user's data can be recovered using maximum likelihood detection irrespective of cor-relation values (q). For highly correlated channels, utilizing constellation rearrangement of the transmitted signals enables maximum achievable capacity by resolving the possible ambiguity of signal detection. For example, it is shown for four users employing QPSK modulation and communicating with three antenna receiver that maximum 8 bit/s/Hz can be achieved using UPSC and RSC compared with 4.64 bit/s/Hz for ISC at extreme correlation level of q = 1. The superiority of these methods provides a potential step towards efficient utilization of the available spectrum for next generation wireless and mobile com- munications in different channel environments.

Introduction

Efficient exploitation of the available spectrum is one of most challenging tasks to meet the increasing demands of wireless and mobile communication services such as internet and media rich applications. Therefore, different promising multiple access techniques are considered to fulfil the high spectral efficiency requirements of next generation systems. The most important schemes are represented by the hybrid combination of multiple-input Multiple-output (MIMO) technology with orthogonal frequency division multiplexing (OFDM), orthogonal frequency division multiple access (OFDMA), and multi-carrier code division multiple access (MC-CDMA). Recently, in fading signature multiple access (FSMA), multiple users can simultaneously transmit their signals without subdivision in orthogonal dimensions of time, frequency, space and codes by exploiting the channel difference as a multiple access dimension. In this scheme, space diversity using multiple receive antenna can be employed to enhance the system performance. However, for capacity evaluation in real propagation environment, channel correlation has direct impact rather than the basic assumption of independent fading among users. As the correlation increased, the channel capacity is degraded significantly, Many research studies have dealt with the correlation and its effect on the channel capacity such as. For example, high antenna correlations may severely degrade the capacity of MIMO communication systems to that of a single-input multiple-output (SIMO), multiple-input single-output (MISO), single-input single-Output (SISO) systems. Also, despite of the natural spatial difference among users in multiuser MIMO (MU-MIMO) systems, capacity degradation may occur.

To mitigate the correlation effects even for extreme level of correlation, many techniques have been proposed such as the key element is not to depend on the fading channel difference alone for signal detection. For example, in a precoding technique based on phase and power adjustment of the input constellations has been introduced for SU-MIMO with transmits correlation. However, this method assumes the availability of correlation properties at the transmitter side and cannot be applied directly for multiuser case without proper modifications. Therefore, different scenarios have been presented in for signal design of two BPSK users equipped with a single antenna by applying constellation rearrangement to increase the minimum Euclidean distance (d_{min}) of the composite received signals. Receive antenna selection is utilized for the uplink correlated MU-MIMO to maximize the channel and user capacity.

Although the idea of signal constellation rearrangement seems old fashioned it is still efficient and promising approach to solve various problems in different wireless schemes and applications. For example, it has been adopted in recent works to enlarge the capacity of multiple access channels and wireless relay systems with network coding In other previous studies, it has been used to improve the transmit diversity and error performance of wireless communication systems. They are based on maximizing d_{min} of the composite receive signal by rearranging the signal constellation of individual mobile users. At the receive side, maximum likelihood detection (MLD) is employed to recover the users' data irrespective of correlation level. Performance and channel capacity evaluations of MIMO systems over Rayleigh fading environment have been investigated in many researches [and for different schemes such as spread spectrum antenna selection, OFDM, and

OFDMA. In particular, channel capacity is a major performance measure which indicates the maximal data rate supported by the communication system with error-free transmission. Therefore, it is derived in this study for FSMA based on the well known Ungerboeck's capacity expression The effectiveness of proposed methods is validated through computer simulations under various channel correlations q in the interval ½0; 1& compared with the conventional identical signal constellation (ISC) sce-nario. This will enable efficient exploitation of the limited spectral resources in modern and future wireless and mobile communications over different channel conditions.

Applications

Fading signature multiple access (FSMA) is a promising radio access technology that utilizes the difference in users' fading channels for joint signal detection. However, channels' correlation has direct impact to degrade the capacity of wireless communications considerably. In this paper, efficient signal design methods are proposed to mitigate the correlation effects and maintain highest achievable channel capacity of correlated FSMA system. Two methods named as unequal power signal constellation (UPSC) and rotated signal constellation (RSC) are considered and compared with the conventional identical signal constellation (ISC) technique. Based on the principles of maximizing the minimum Euclidean distance of composite received signals, user's data can be recovered using maximum likelihood detection irrespective of correlation values (ρ). For highly correlated channels, utilizing constellation rearrangement of the transmitted signals enables maximum achievable capacity by resolving the possible ambiguity of signal detection. For example, it is shown for four users employing QPSK modulation and communicating with three antenna receiver that maximum 8 bit/s/Hz can be achieved using UPSC and RSC compared with 4.64 bit/s/Hz for ISC at extreme correlation level of $\rho = 1$. The superiority of these methods provides a potential step towards efficient utilization of the available spectrum for next generation wireless and mobile communications in different channel environments.

a. UPSC and RSC are designed by exploiting the advan-tages of spatial difference among mobile users, power control at base station (BS), and channel state infor-mation (CSI) to mitigate the effects of channel correlation on FSMA scheme.

b. Transmit signal design of each user is selected based on maximum d_{min} of composite receive signal constella-tion through power control and/or partial feedback of channel phase.
c. Additional enhancement to the users' channel differ-ence is achieved by maximizing d_{min} of received constellation through proposed methods. This leads to resolve the inherent detection ambiguity in correlated channels and enables robust solution for high spectral efficiency.
d. Constellation constrained capacity formula for FSMA scheme is derived to evaluate the system performance and show the effectiveness of proposed designs com-pared with ISC technique.

Busy Tone Multiple Access Protocol (BSMA)

Multichannel protocol where the channel is separated into:

1. control channel: used for data transmission;
2. Data channel: used for busy tone transmission.

BTMA works as follows:

A node senses the control channel to check whether the busy tone is active

If not, it turns on the busy tone signal and starts data transmission

If yes, the node waits for a random period of time and repeats the procedure

Any node that senses the carrier on the incoming data channel also transmits a busy tone

BTr (set up by receiver) notifies

- the RTS sender that RTS has been received and channel has been acquired
- Announces to its neighbor nodes that it is receiving data packet and they should refrain from accessing the channel

BTt (set up by sender)

- Provides protection for the RTS packet

4.15. DBTMA Protocol

An ad hoc network is a collection of wireless hosts forming a temporary network without relying on an established infrastructure or on a central control. Network operations, such as routing, are performed in a distributed and cooperative manner. The applications of ad hoc networks are in situations in which the network needs to be deployed rapidly, such as communications in emergency situations. Recently, the Bluetooth technology was introduced the Bluetooth devices can form an ad hoc network to communicate with each other. Due to the large span of ad hoc networks and limited radio transmission range, multi-hop routing is usually used, in which the communication between any two nodes is performed by forwarding the data packet from one node to another until the packet reaches the destination.

DBTMA Protocol

In the DBTMA protocol, two narrow-bandwidth tones are implemented with enough spectral separation on the single shared channel. BT_t (the transmit busy tone) and BT_r (the receive busy tone), indicate whether the node is transmitting RTS packets or receiving data packets, respectively. The transmit busy tone (BT_t) provides protection for the RTS packets to increase the probability of successful RTS reception at the intended receiver. Use the receive busy tone (BT_r) to acknowledge the RTS packet and provide continuous protection for the transmitted data packets. All nodes sensing any busy tone are not allowed to send RTS requests. When the start of the signal is sensed, a node sending the RTS packet is required to abort such transmission immediately. Indeed, the RTS packets and the receive busy tone solve the hidden- and the exposed-terminal problems.

Assumptions on DBTMA

- The radio transmission range of the ad hoc network in which the DBTMA scheme operates is on the order of tens to hundreds of meters. There is no capture effect or fading on the channel.

- Any overlap of transmissions at a receiver causes the receiver not to understand either packet. Packet collisions are the only source of packet errors.
- The data processing time and the transmit/receive turnaround time at each node are negligible.
- The busy tone signal and the data signal have the same transmission range.
- The interference between the busy tone signals and the data signal is negligible.

Summary

- *The DSDV update message consists of three fields: Destination Address, Sequence Number and Hop Count.*
- *Destination-Sequenced Distance-Vector Routing (DSDV) is a table-driven routing scheme for ad hoc mobile networks based on the Bellman–Ford algorithm.*
- *The AODV routing protocol is designed for mobile ad hoc networks with populations of tens of thousands of mobile nodes*
- *The DRP is responsible for maintaining what is called the Signal Stability Table and also the Routing Table.*
- *The Optimized Link State Routing Protocol (OLSR) is an IP routing protocol optimized for mobile ad hoc networks, which can also be used on other wireless ad hoc networks.*
- *The LMR algorithm is a link reversal algorithm which addresses the issue of partitioned networks found with the Gafni-Bertsekas algorithms, by providing a link erasure mechanism.*
- *LBAR defines a new metric for routing known as the degree of nodal activity to represent the load on a mobile node.*
- *Delay-based Load-Aware On-demand Routing (D-LAOR) protocol determines the optimal path based on the estimated total path delay and the hop count.*

Chapter V

Energy Conservation Schemes in Wireless Networks

Objective

- *To learn about the fundamentals of Sensor Networks*
- *To know the classifications of energy conservation schemes*
- *To become familiar with the concepts of duty cycling, sleep/wakeup and data driven approaches*
- *To understand the application of In-network processing*
- *To interpret the concepts of Mobility based approaches*

5.0. Introduction

A wireless sensor network consists of sensor nodes deployed over a geographical area for monitoring physical phenomena like temperature, humidity, vibrations, seismic events, and so on. Typically, a sensor node is a minute device that consists of three components such as a sensing subsystem for data attainment from the physical surrounding environment, a processing subsystem for local data processing and storage, and a wireless communication subsystem for data transmission. In addition, an energy source supplies the energy needed by the device to perform the planned task. Energy consumption is one of the biggest constraints of the wireless sensor node and this limitation combined with a typical deployment of large number of nodes has added many challenges to the design and management of wireless sensor networks.

They are typically used for remote environment monitoring in areas where providing electrical power is difficult. Therefore, the devices need to be powered by batteries and alternative energy sources. Because battery energy is limited, the use of different techniques for energy saving is one of the hottest topics in WSNs. This energy source habitually consists of a battery with a limited energy resource. In addition, it could be inconvenient to recharge the battery, because nodes may be deployed in a hostile or unpractical environment.

On the other hand, the sensor network should have a lifetime long enough to fulfill the application requirements. In many cases, a lifetime in the order of several months, or even years, may be required. In some cases, it is possible to scavenge energy from the external environment (e.g., by using solar cells as energy source). However, external energy supply sources often display a non-continuous behavior so that an energy buffer (a battery) is needed as well. In any case, energy is a very critical resource and must be used very sparingly. Therefore, energy conservation is a key issue in the design of systems based on wireless sensor networks.

The main components of a sensor node are microcontroller, transceiver, external memory, energy source and one or more sensors. Microcontroller performs tasks, processes data and controls the functionality of other components in the sensor node. Sensors are used to sense the data from the physical environment, memory is for storage, and a transceiver is used for data transmission.

5.1. Energy Conservation Schemes

Sensor networks are deployed in an ad hoc fashion, with individual nodes remaining largely inactive for long periods of time, but then becoming suddenly active when something is detected. Sensor Networks are generally battery constrained. Prior to discussing the advanced classification of energy conservation proposals, it is very significant to present the network-level and node-level architectures. Obviously, the energy breakdown heavily depends on the specific node as given below:

- The communication subsystem has much higher energy consumption than the computation subsystem. It has been shown that transmitting one bit may consume as much as executing a few thousands instructions. Therefore, communication should be traded for computation.

- The radio energy utilization is of the same order in the reception, transmission, and idle states, while the energy spending drops off at least one order of magnitude in the sleep state. Therefore, the radio should be put to sleep (or turned off) whenever possible.
- Depending on the specific application, the sensing subsystem might be another major source of energy consumption, so its energy consumption has to be reduced as well.

Based on the above architecture and energy breakdown, numerous approaches have to be exploited, at the same time, to reduce energy consumption in wireless sensor networks. At a very common level, three main enabling techniques, namely duty cycling, data-driven approaches and mobility are identified.

Energy Conservation Schemes

- **Duty cycling**
  - Topology Control
    - Location-driven
    - Connection-driven
  - Sleep/wakeup protocols
    - On-demand
    - Scheduled rendezvous
    - Asynchronous
  - MAC (Media Access Control) protocols with low duty cycle
    - TDMA (Time Division Multiple Access)
    - Contention based
    - Hybrid
- **Data-driven**
  - Data reduction
    - In-network processing
    - Data compression
    - Data prediction
      - Stochastic Approaches

        - Time Series Forecasting
        - Algorithmic Approaches
    - Energy –Efficient data Acquisition
        - Adaptive Sampling
        - Hierarchical Sampling
        - Model-based Sampling
- **Mobility based**
    - Mobile-sink
    - Mobile-relay

5.2. Duty Cycling

Duty cycling is primarily focused on the networking subsystem. The most efficient energy-conserving operation is putting the radio transceiver in the (low-energy) sleep mode whenever communication is not required. Ideally, the radio should be switched off as soon as there is no more data to send/receive, and it should be resumed as soon as a new data packet is prepared. In this way, nodes interchange between active and sleep periods depending on network activity. This activity is generally referred to as duty cycling and duty cycle is defined as the part of time when nodes are active during their lifetime.

A common approach for conserving channel capacity and energy is optimizing power-aware routing and different kind of duty cycling (DC) and harvesting technology. As the sensor nodes execute a cooperative task, they need to coordinate their sleep/wake up times. A sleep/wake up scheduling algorithm thus accompanies any duty cycling scheme. It is typically a distributed algorithm based on which sensor nodes decide when to change from active to sleep and back. It allows neighbouring nodes to be active at the same time, thus making packet exchange possible even when nodes operate with a low duty cycle (i.e., they sleep for most of the time).

Duty-cycling schemes are normally unaware of data that are sampled by sensor nodes. In fact, data sensing impacts on sensor nodes' energy consumption in two ways:

- **Extra samples**. As the sampled data generally has strong spatial and/or temporal relationship, there is no need to communicate the redundant information to the sink.

- **Energy consumption of the sensing subsystem**. Reducing communication is not enough when the sensor itself is energy starving.

In the first case, extra samples result in useless energy consumption. Even if the cost of sampling is small, it results in extra communications. The second problem arises every time the consumption of the sensing subsystem is not small.

5.2.1. Topology Control Protocols

The idea of topology control is strictly related to that of network redundancy. Dense sensor networks usually have some degree of redundancy. Energy conservation is a very critical issue in WSN. A lot of work has been done on the techniques of topology control so that sensor nodes which are not in direct use can be put to a low power consuming state. In many cases, network deployment is done at random, e.g., by dropping a large number of sensor nodes from an airplane. Therefore, it may be convenient to deploy a number of nodes greater than necessary to cope up with the possible node failures occurring during or after the deployment. In various contexts, it is much easier to deploy at first a greater number of nodes than re-deploying additional nodes when needed.

A number of criteria can be used to choose which nodes are to be activated/deactivated and when. In this view, topology control protocols can be broadly classified into two categories. One is location driven protocols that describe which node should be turned on and when, based on the location of sensor nodes which are assumed to be known. Another is the connectivity driven protocols which dynamically activate/deactivate sensor nodes in such a way that network connectivity or complete sensing coverage is fulfilled.

5.2.2. Location-Driven

GAF (Geographical Adaptive Fidelity) is a location-driven protocol that reduces energy consumption while keeping a constant level of routing fidelity. The sensing area where nodes are distributed is divided into small virtual grids. All nodes within the same virtual grid are equivalent for routing and just one node at a time needs to be active. Therefore, nodes have to coordinate with each other to decide which one can sleep and how long.

Initially, a node starts in the discovery state where it exchanges discovery messages with other nodes. After broadcasting the message, the node enters the active state. While active, it periodically re-broadcasts its discovery message. A node in the discovery or active state can change its state to sleep when it detects that some other equivalent node will handle routing. Nodes in the sleeping state wake up after a sleeping time and go back to the discovery state. In GAF, load balancing is achieved through a periodic re-election of the leader, i.e., the node will remain active to manage routing in the virtual grid. The leader is chosen through a rank-based election algorithm which considers the nodes' residual energy, thus allowing the network lifetime to increase in proportion to node density.

GAF is independent of the routing protocol, so that it can be used along with any existing solution of that kind. In addition, GAF does not significantly affect the performance of the routing protocol in terms of packet loss and message latency. However, the structure imposed over the network may lead to an underutilization of the radio coverage areas. In fact, as all nodes within a virtual grid must be able to reach any node in an adjacent virtual grid, the nodes are actually forced to cover less than half the distance allowed by the radio range.

Although being defined as a geographic routing protocol, GeRaF (Geographic Random Forwarding) actually presents features which are in the direction of location-driven duty-cycled operations. They make use of both nodes position and redundancy. Nodes follow a given duty cycle to switch between awake (active) and sleep (inactive) states. Nodes periodically switch to the active state, starting with a listening time, so that they can participate in routing if needed. Data forwarding starts as soon as a node has a packet to send. In this case, the node becomes active and broadcasts a packet containing its own location and the location of the intended receiver. Then a receiver-initiated forwarding phase takes place. As a result, one of the active neighbors of the sender will be selected to relay the packet towards the destination.

The main idea is that each active node has a priority which depends on its closeness to the intended destination of the packet. In addition to priority, a distributed randomization scheme is also used, in order to reduce the probability that many neighboring nodes are simultaneously sleeping. Specifically, the portion of the coverage area of the sender which is closer to the intended destination is split into a number of regions. Each region has its associated priority and regions are chosen so that all the nodes within a region are closer to the destination than any other node in a region with a lower priority.

After the broadcast, nodes in the region with the higher priority contend for forwarding. If only one node gets the channel, it simply forwards the packet and the process ends. Otherwise, multiple nodes may transmit simultaneously, resulting in a collision. In this case, a resolution technique is applied in order to select a single forwarder. There may also be the case in which no node can forward the packet because all nodes in the region are sleeping. In the next transmission attempt, the forwarder will be chosen among nodes in the second highest-priority region and so on. Every time the relay selection phase will be repeated until a maximum number of retries will be reached. Eventually, after a hop-by-hop forwarding, the packet will reach the intended destination.

5.2.3. Connectivity-Driven

Span is a connectivity-driven protocol that adaptively elects "coordinators" of all nodes in the network. Coordinators stay awake continuously and perform multi-hop routing, while the other nodes stay in sleeping mode and periodically check if it is needed to wake up and become a coordinator. To guarantee a sufficient number of coordinators Span uses the following coordinator eligibility rule: if two neighbors of a non-coordinator node cannot reach each other, either directly or via one or more coordinators, that node should become a coordinator. However, it may happen that several nodes discover the lack of a coordinator at the same time and thus, they all decide to become a coordinator. To avoid such cases, nodes that decide to become a coordinator defer their announcement by a random backoff delay.

Each node uses a function that generates a random time by taking into account both the number of neighbors that can be connected by a potential coordinator node and its residual energy. The fundamental ideas are that (i) nodes with a higher expected lifetime should be more likely to volunteer to become a coordinator and (ii) coordinators should be selected in such a way as to minimize their number. Each coordinator periodically checks if it can stop being a coordinator. In detail, a node should withdraw as a coordinator if every pair of its neighbors can communicate directly or through some other coordinators.

To avoid loss of connectivity, during the transient phase, the old coordinator continues its service until the new one is available. The Span election algorithm requires knowing the neighbor and connectivity information to decide whether a node should become a coordinator or not. Such information is provided by the routing protocol. Hence span depends on it and it may require modification in the routing look up process.

ASCENT (Adaptive Self-Configuring SEnsor Networks Topologies) is a connectivity-driven protocol that, unlike span, does not depend on the routing protocol. In ASCENT, a node decides whether to join the network or continue to sleep based on information about connectivity and packet loss that are measured locally by the node itself. The basic idea of ASCENT is that initially only some nodes are active, while all other ones are passive, i.e., they listen to packets but do not transmit. If the number of active nodes is not large enough, the sink node may experience a high message loss from sources.

The sink then starts sending help messages to solicit neighboring nodes that are in the passive state (passive neighbors) to join the network by changing their state from passive to active (active neighbors). Passive neighbors have their radio on and listen to all packets transmitted by their active neighbors. However, they do not cooperate to forward data packets or exchange routing control information they only collect information about the network status without interfering with other nodes. On the contrary, active neighbors forward data and routing (control) messages until they run out of energy.

Active nodes can also send help messages when they find the local data loss at an unacceptable level. As soon as it joins the network, a node starts monitoring the network conditions and also signals its presence as an active node through a neighbor announcement message. This process continues until the number of active nodes is such that the message loss rate experienced by the sink is below a pre-defined application-dependent threshold.

The process will re-start when some future network event (e.g., a node failure) or a change in the environmental conditions causes an increase in the message loss. As mentioned above, ASCENT is independent of the routing protocol. In addition, it limits the packet loss due to collisions because the nodes density is explicitly taken into account as a parameter (in the form of a neighbor threshold value).

5.3. SLEEP/WAKE UP PROTOCOLS

As previously discussed, sleep/wake up schemes can be defined for a given component (i.e., the radio subsystem) of the sensor node, without relying on topology or connectivity aspects. In this section, a survey of the main sleep/wake up schemes which are implemented as independent protocols on top of the MAC protocol (i.e., at the network or the application layer) has been carried out. Independent sleep/wake up protocols can be further subdivided into three main categories: on-demand scheduled rendezvous and asynchronous schemes.

5.3.1. On-Demand

On-demand protocols take the most intuitive approach to energy management. The basic idea is that a node should wake up only when another node wants to communicate with it. The main problem associated with on-demand schemes is how to inform the sleeping node that some other node is willing to communicate with it.

Actually, the approach taken by on-demand protocols is the ideal one, because it maximizes energy saving as nodes remain active only for the minimum time required for communication. In addition, there is only a very limited impact on latency because the target node wakes up immediately as soon as it realizes that there is a pending message. Unfortunately, the adoption of a radio triggered wake up scheme is almost always impractical, since it can only be applied when the distance between nodes is very short indeed (a few meters). Hence, introducing an additional wakeup radio is a more promising direction, especially suitable to event detection applications.

However, the wake up radio is costly and generally it is not shipped with commonly used sensor platforms. So when a second radio is not available or convenient, other solutions such as the scheduled rendezvous and the asynchronous wake up schemes can be used. Both of them trade energy savings for an increased latency experienced by messages to travel through several hops.

5.3.2. Scheduled Rendezvous

An alternative solution is using a scheduled rendezvous approach. The basic idea behind scheduled rendezvous schemes is that each node should wake up at the same time as its neighbours. Typically, nodes wake up according to a wake up schedule and remain active for a short time interval to communicate with their neighbours. Then, they go to sleep until the next rendezvous time.

The scheduled rendezvous approach is convenient because it is suitable to data aggregation and supports broadcast traffic. Unfortunately, it requires nodes to be synchronized, in which some cases can be difficult to achieve or expensive in terms of additional protocol overhead for synchronization. On the other hand, asynchronous wake up protocols do not need a tight synchronization among network nodes. In addition, asynchronous schemes are generally easier to implement and can ensure network connectivity even in highly dynamic scenarios where synchronous (i.e., scheduled rendezvous) schemes become inadequate. This greater flexibility is compensated by lower energy efficiency.

5.3.3. Asynchronous

Finally, an asynchronous sleep/wake up protocol can be used. With these protocols, a node can wake up when it wants and still be able to communicate with its neighbours. This goal is achieved by properties implied in the sleep/wake up scheme. Thus no explicit information exchange is needed among the nodes.

In the asynchronous schemes, nodes need to wake up more frequently than in scheduled rendezvous protocols. Therefore, asynchronous protocols usually result in a higher duty cycle for network nodes than their synchronous counterparts. In addition, the support to broadcast traffic is problematic. Due to their wider applicability and their properties, scheduled rendezvous and asynchronous approaches seem to be the most promising solutions in the class of sleep/wake up protocols.

5.4. MAC PROTOCOLS WITH LOW DUTY CYCLE

The tremendous and rapid development in sensors technology allowed their application in various fields requiring monitoring, such as, transportations, rare species surveillance, agriculture, military activities, medical field, etc. Due to their intrinsic constraints and limitations, several dedicated MAC protocols have been designed for wireless sensor networks and whose main objectives are bandwidth optimization while keeping very low energy consumption

In the physical layer, the use of directive and adaptive phased arrays is proposed for the WSN gateways, increasing the communication range between sensors and their gateway. In wireless communication, idle listening, receiving and transmitting are the main source of consumption of energy. The most common MAC protocols are TDMA (Time Division Multiple Access) based, contention-based, and hybrid.

5.4.1. TDMA

TDMA (Time Division Multiple Access) schemes naturally enable a duty cycle on sensor nodes as channel access is done on a slot-by-slot basis. As nodes need to turn on their radio only during their own slots, the energy consumption is ideally reduced to the minimum level required for transmitting/receiving data. In TDMA-based MAC protocols, time is divided into (periodic) frames and each frame consists of a certain number of time slots. Every node is assigned to one or more slots per frame, according to a certain scheduling algorithm, and it uses such slots for transmitting/receiving packets to/from other nodes.

In many cases, nodes are grouped to form clusters with a cluster-head which is in charge of assigning slots to nodes in the cluster, LEACH and Energy-aware TDMA-based MAC. One of the most important energy efficient TDMA protocol for wireless sensor networks is TRAMA. TRAMA divides time into two portions, a random-access period and a scheduled access period. The random access period is devoted to slot reservation and is accessed with a contention-based protocol. On the contrary, the scheduled access period is formed by a number of slots assigned to an individual node.

The slot reservation algorithm is as follows. First, nodes that derive two-hop neighborhood information are required to establish collision free schedules. Then, nodes start an election procedure to associate with each slot having a single node. Every node gets the priority of being the owner of a specific slot. This priority is calculated as a hash function of the node identifier and the slot number. The node with the highest priority becomes the owner of a given slot.

Finally, nodes send out a synch packet containing a list of intended neighbor destinations for subsequent transmissions. As a consequence, nodes can agree on the slots which they must be awake in. Unused slots can be advertised by their owners for being re-used by other nodes.

FLAMA (FLow-Aware Medium Access) is a TDMA MAC protocol derived from TRAMA and optimized for periodic monitoring applications. The main idea is to avoid the overhead associated with the exchange of traffic information. As the message flow in periodic reporting applications is rather stable, FLAMA first sets up flows and then uses a pull-based mechanism, so that the data are transferred only after being explicitly requested.

5.4.2. Contention-Based Protocols

Contention-based protocols are the most popular class of MAC protocols for wireless sensor networks. They achieve duty cycling by tightly integrating channel access functionalities with a sleep/wake up scheme similar to those described above. The only difference is that in this case the sleep/wake up algorithm is not a protocol independent of the MAC protocol, but is tightly coupled with it.

Most of MAC protocols proposed for wireless sensor networks are contention-based protocols. One of the most popular contention-based MAC protocols is B-MAC (Berkeley MAC), a low complexity and low power MAC protocol which is shipped with the TinyOS operating system. The goal of B-MAC is to provide a few core functionalities and an energy efficient mechanism for channel access. First, B-MAC implements basic channel access control features: a backoff scheme, an accurate channel estimation facility and optional acknowledgements.

Second, to achieve a low duty cycle B-MAC uses an asynchronous sleep/wake scheme based on periodic listening called Low Power Listening (LPL). Nodes periodically wake up to check the channel for activity. The period between consecutive wakeups is called check interval. After waking up, nodes remain active for a wake up time, in order to properly detect eventual ongoing transmissions. While the wake up time is fixed, the check interval can be specified by the application.

B-MAC packets are made up of a long preamble and a pay load. The preamble duration is at least equal to the check interval so that each node can always detect an ongoing transmission during its check interval. This approach does not require nodes to be synchronized. In fact, when a node detects channel activity, it just remains active and receives first the preamble and then the payload.

A well-known MAC protocol for multi-hop sensor networks is S-MAC (Sensor-MAC), which adopts a scheduled rendezvous communication scheme. Nodes exchange sync packets to coordinate their sleep/wake up periods. Every node can establish its own schedule or follow the schedule of a neighbor by means of a random distributed algorithm. Nodes using the same schedule form a virtual cluster. A node can eventually follow both schedules if they do not overlap, so that it can bridge communication between different virtual clusters.

The channel access time is split into two parts. In the 'listen' period nodes exchange sync packets and special control packets for collision avoidance. In the remaining period, the actual data transfer takes place. The sender and the destination node are awake and talk to each other. Nodes not concerned with the communication process can sleep until the next listen period.

To avoid high latencies in multi-hop environments S-MAC uses an adaptive listening scheme. A node overhearing its neighbor's transmissions wakes up at the end of the transmission for a short period of time. If the node is the next hop of the transmitter, the neighbor can send the packet to it without waiting for the next schedule. The parameters of the protocol, i.e., the 'listen and the sleep' period, are constants and cannot be varied after the deployment.

5.4.3. Hybrid Protocols

Hybrid protocols adapt the protocol behavior to the level of contention in the network. They behave as contention-based protocols when the level of contention is low and switch to a TDMA scheme when the level of contention is high. The basic idea behind hybrid MAC protocols is switching the protocol behavior between TDMA and CSMA, depends on the level of contention and this is not new.

In PTDMA, time is slotted and nodes are distinguished as owners and non-owners. The protocol adjusts the access probability of owners and non-owners depending on the number of senders. By doing so, it adapts the MAC protocol to work as a TDMA or CSMA scheme depending on the level of contention in the network. However, PTDMA was conceived for a one-hop wireless scenario. Therefore, it does not take into account issues such as topology changes, synchronization errors, interference irregularities which are quite common in wireless sensor networks.

In the specific context of wireless sensor networks, one of the most interesting hybrid protocols is Z-MAC. In order to define the main transmission control scheme, Z-MAC starts a preliminary set up phase. By means of the neighbor's discovery process each node builds a list of two-hop neighbors. Then a distributed slot assignment algorithm is applied to ensure that any two nodes in the two-hop neighborhood are not assigned to the same slot.

As a result, it is guaranteed that no transmission from a node to any of its one-hop neighbor interferes with any transmission from its two-hop neighbors. The local frame exchange is aimed at deciding the time frame. Z-MAC does not use a global frame equal to all nodes in the network. It would be very difficult and expensive to adapt when a topology change occurs. Instead, Z-MAC allows each node to maintain its own local time frame that depends on the number of neighbors and avoids any conflict with its contending neighbors.

The local slot assignment and time frame of each node are then forwarded to its two-hop neighbors. Thus any node has slot and frame information about any two-hop neighbors and all these synchronize to a common reference slot.

5.5. Data-Driven Approaches

Data driven techniques presented below are designed to reduce the amount of sampled data by keeping the sensing precision within an acceptable level for the application. Data-driven approaches can be separated according to the difficulty they address. In particular, data-reduction schemes address the case of unwanted samples, as energy-efficient data acquisition schemes are mainly aimed at reducing the energy spent by the sensing subsystem.

5.5.1. Data Reduction

It is important to discuss here one more classification level related to data-reduction schemes. These techniques are in-network processing, data compression and data prediction. All these techniques aim at reducing the quantity of data to be delivered to the sink node. Scenarios where nodes have limited energy and forward messages of different priorities are frequent in the context of wireless sensor networks. Tailored to those scenarios, this relies on stochastic tools to develop selective message forwarding schemes.

5.5.2. In-Network Processing

In-network processing consists of performing data aggregation (e.g., computing average of some values) at intermediate nodes between the sources and the sink. In this way, the amount of data is reduced while traversing the network towards the sink. The most suitable in-network processing technique depends on the specific application and it must be modified accordingly.

5.5.3. Data Compression

Data compression can be applied to reduce the amount of information sent by source nodes. This scheme involves encoding information at nodes which generate data and decode it at the sink.

5.5.4. Data Prediction

Data prediction involves building an abstraction of a sensed phenomenon, i.e., a model describing data evolution. The model can expect the values sensed by sensor nodes within certain error limits and they reside both at the sensors and at the sink. If the required accuracy is fulfilled, queries issued by users can be evaluated at the sink through the model without the need to get the accurate data from the nodes.

There are two instances of a model in the network, one residing at the sink and the other at source nodes (so that there are as many pairs of models as sources). The model at the sink can be used to answer queries without requiring any communication, thus reducing the energy consumption. Clearly, this operation can be performed only if the model is a valid representation of the phenomenon at a given instant.

Sensor nodes take data samples as usual and compare the actual data against the prediction. If the sensed value falls within an application-dependent tolerance, then the model is considered valid. Otherwise, the source node may transmit the sampled data and/or start a model update procedure involving the sink as well. The features of a specific data prediction technique depend on the way the model is built. Data prediction techniques can be split into three main classes:

(i) Stochastic approaches
(ii) Time series forecasting
(iii) Algorithmic approaches

Techniques belonging to the first class derive a stochastic characterization of the phenomenon, i.e., in terms of probabilities and/or statistical properties. Two main approaches of this kind are as follows. On the one hand, it is possible to map data into a random process described in terms of a probability density function. Data prediction is then obtained by combining the computed probability density function with the observed samples. On the other hand, a state space representation of the phenomenon can be derived, such that forthcoming samples can be guessed by filtering out a non-predictable component modeled as noise.

The second class of data prediction techniques is time series forecasting, where a set of historical values (the time series) obtained by periodical samplings is used to predict a future value in the same series. Generally, a time series can be represented as a combination of a pattern and a random error. The pattern, in turn, is characterized by its trend, i.e., its long-term variation and its seasonality, i.e., its periodical fluctuation.

Finally, the last class of data prediction techniques relies on a heuristic or a state-transition model describing the sensed phenomenon. Such algorithmic approaches derive methods or procedures to build and update the model on the basis of the chosen characterization.

5.5.5. Energy Efficient Data Acquisition

An emerging class of applications is actually sensing-constrained. This is in contrast with the general assumption that sensing is not relevant from an energy consumption standpoint. In fact, the energy consumption of the sensing subsystem not only may be relevant, but it can also be greater than the energy consumption of the radio or even greater than the energy consumption of the rest of the sensor nodes. This can be due to many different factors as given below:

- Energy hungry transducers. Some sensors intrinsically require high energy resources to perform their sampling task.
- Active sensors. Another class of sensors can get data about the sensed phenomenon by using active transducers (e.g., sonar, radar or laser rangers). In this case, sensors have to send out a probing signal in order to acquire information about the observed quantity. Long acquisition time. The acquisition time may be in the order of hundreds of milliseconds or even seconds; hence the energy consumed by the sensing subsystem may be high, even if the sensor energy consumption is moderate.

In this case, reducing communications may not be enough, but energy conservation schemes have to actually reduce the number of acquisitions (i.e., data samples). It should also be pointed out that energy efficient data acquisition techniques are not exclusively aimed at reducing the energy consumption of the sensing subsystem. By reducing the data sampled by source nodes, they decrease the number of communications as well. Actually, many energy-efficient data-acquisition techniques have been conceived for minimizing the radio energy consumption, under the assumption that the sensor consumption is negligible.

(i) Adaptive sampling
(ii) Hierarchical sampling
(iii) Model-based active sampling

Adaptive Sampling

Adaptive sampling techniques exploit such similarities to reduce the amount of data to be acquired from the transducer.

Hierarchical Sampling

The hierarchical sampling approach assumes that nodes are equipped with different types of sensors. As each sensor is characterized by a given resolution and its associated energy consumption, this technique dynamically selects which class needs be activated, in order to get a tradeoff between accuracy and energy conservation.

Model –Based Active Sampling

Model-based active sampling takes an approach similar to data prediction. A model of the sensed phenomenon is built upon sampled data, so that future values can be forecast with certain accuracy.

5.6. Mobility-Based Schemes

In case, some of the sensor nodes are mobile, mobility can finally be used as a tool for reducing energy consumption (beyond duty cycling and data-driven techniques). In a static sensor network, packets coming from sensor nodes track a multi-hop path towards the sink (s). Thus, a few paths can be more loaded than others and nodes closer to the sink have to relay more packets so that they are more subject to early energy reduction. If some of the nodes (including, possibly, the sink) are mobile, the traffic flow can be changed if mobile devices are responsible for data collection directly from static nodes.

Normal nodes wait for the channel of the mobile device and route messages towards it, so that the communications take place in proximity (directly or at most with a limited multi-hop traversal). As a result, ordinary nodes can save energy because path length, contention and forwarding overheads are reduced as well. In addition, the mobile device can visit the network in order to extend more evenly the energy consumption due to communications. When the cost of mobilizing sensor nodes is excessive, the normal approach is to attach sensor nodes to entities that will be roaming in the sensing field anyway, such as buses or animals.

Mobility-based schemes can be classified as mobile-sink and mobile-relay schemes, depending on the type of the mobile entity. It is significant to point out here that when considering mobile schemes, an important issue is the type of control the sensor-network designer has on the mobility of nodes. Mobile nodes can be separated into two broad categories: they can be specifically designed as part of the network infrastructure or they can be part of the environment. When they are part of the infrastructure, their mobility can be fully controlled and are, in general, robotized.

When mobile nodes are part of the environment they might not be controllable. If they follow a strict schedule, then they have a fully predictable mobility. If not, they may have a random behavior so that no reliable assumption can be made on their mobility. In such a case, mobility patterns can be learned based on successive observations and estimated with some accuracy.

For example, sensors can be equipped with mobilizers for changing their location. As mobilizers are generally quite expensive from the energy consumption standpoint, adding mobility to sensor nodes may not be convenient. In fact, the resulting energy consumption may be greater than the energy gain due to the mobility itself.

So, instead of making each sensor node mobile, mobility can be limited to special nodes which are less energy constrained than the ordinary ones. In this case, mobility is strictly tied to the heterogeneity of sensor nodes. On the other hand, instead of providing mobilizers, sensors can be placed on elements which are mobile on their own (e.g., animals, cars and so on).

There are two different options in this case. First, all the sensors are put onto mobile elements, so that all nodes in the network are mobile. Alternatively, only a limited number of special nodes can be placed on mobile elements, while the other sensors are stationary. Anyway, in both cases there is no additional energy consumption overhead due to mobility, but the mobility pattern of mobile elements has to be taken into account during the network design phase (more details are provided below). As mentioned in the previous section, mobility-based energy conservation schemes can be classified depending on the nature of the mobile element, i.e., a Mobile Sink (MS) or a Mobile Relay (MR).

5.6.1. Mobile-Sink-Based Approaches

Many approaches proposed in the literature about sensor networks with mobile sinks (MSs) rely on a Linear Programming (LP) formulation which is exploited in order to optimize parameters such as network lifetime and so on. During visits to nodes, the sink stays at the node location for a period of time. Nodes not in the coverage area of the sink can send messages along multi-hop paths ending at the MS and obtained using the shortest path routing.

5.6.2. Mobile-Relay-Based Approaches

In mobile wireless sensor networks (WSN) coverage can be enhanced by moving the sensors so that a better arrangement is achieved. The Mobile Relay (MR) model for data collection in multi-hop sensor networks has already been explored in the context of opportunistic networks. Message ferries are special mobile nodes which are introduced into a sparse mobile ad hoc network to offer the service of message relaying.

Message ferries move around in the network area and collect data from source nodes. They carry stored data and forward them towards the destination node. Thus, message ferries can be seen as a moving communication infrastructure which accommodates data transfer in sparse wireless networks.

In fact, changing the trajectory of the MR is not always possible in the case of sensor networks because sensors may be deployed in places with obstacles, on rough terrain or generally where unmanned vehicles can move only in certain directions. Energy saving is addressed in such a way that a large number of nodes is visited by the MR and it can thus transmit data over a single hop connection using a short range radio.

The other nodes which are not in proximity of the path followed by MR send their data over a multi-hop path. This is however shorter and thus cheaper, with respect to the path established towards a fixed sink node in a classical dense wireless sensor network.

To manage this kind of data collection, nodes self-organize into clusters where cluster heads are the nodes. These nodes are nearer to the path of the MR whereas the other nodes of the cluster send their data to the cluster head for storage until the next visit of the MR. Data from the sensor nodes of the cluster travel towards the cluster heads according to the directed diffusion protocol. Then the election of the cluster heads is kept after the first traversal of the MR.

During this traversal, the MR does not collect any data. Transmissions from cluster heads to the MR occur only when the MR is in proximity so as not to waste energy in useless transmissions. As the trajectory of the MR is assumed to be fixed, it can be controlled only in time. The MR can move at a constant speed worked out, for example, depending on the buffer constraints of the cluster heads. Each cluster head is thus visited before its buffer runs out of space.

However, a better performance is experienced when the MR alternates between two states: moving at a certain constant speed or stopping. So MR moves fast in places with no or only a few sensors and stops in proximity of cluster heads where sensor deployment is denser. The determination of places where sensor deployment is denser (congested regions) is done at each traversal of the MR.

CONCLUSION

- *A wireless sensor network consists of sensor nodes deployed over a geographical area for monitoring physical phenomena like temperature, humidity, vibrations, seismic events, and so on.*
- *The most efficient energy-conserving operation is putting the radio transceiver in the (low-energy) sleep mode whenever communication is not required.*
- *Duty-cycling schemes are normally unaware of data that are sampled by sensor nodes.*
- *GAF is independent of the routing protocol, so that it can be used along with any existing solution of that kind.*
- *Span is a connectivity-driven protocol that adaptively elects "coordinators" of all nodes in the network.*
- *Hybrid protocols adapt the protocol behavior to the level of contention in the network.*
- *Mobility-based schemes can be classified as mobile-sink and mobile-relay schemes, depending on the type of the mobile entity.*

About the Authors

S. Anandamurugan, ME, PhD
Department of Information Technology
School of Communication and Computer Sciences
Kongu Engineering College, Perundurai
Tamilnadu, India

Dr. S. Anandamurugan obtained his Bachelor's degree in Electrical and Electronics Engineering from "Maharaja Engineering College - Avinashi" under Bharathiyar University and Masters Degree in Computer Science and Engineering from "Arulmigu Kalasalingam College of Engineering – Krishnan Koil" under Madurai Kamaraj University. He completed his PhD in Wireless Sensor Networks from Anna University, Chennai. He has 15 years of teaching experience. Currently he is working as an Assistant Professor (Selection Grade) in the department of Information Technology in Kongu Engineering College, Perundurai. He is a life member of ISTE, CSI & ACEEE. He has received "Best Staff" award for the year 2007-08. He has authored more than 70 books. He has Published 30 papers in International and National Journals and 10 Papers in International and National Conferences. His area of interest includes Sensor Networks and Green Computing. He is an Editorial Board Member of the International Journal of Computing Academic Research (IJCAR). He has organized 1CSIR sponsored seminar for the benefit of faculty members and students. He has attended about 40 Seminars, FDP's, and Workshops organized by various Engineering colleges.

P. S. Nandhini, ME
Department of Computer Science and Engineering
School of Communication and Computer Sciences
Kongu Engineering College, Perundurai
Tamilnadu, India

Ms. P. S. Nandhini obtained her Bachelor's degree in Information Technology from "Avinashilingam Institute for Home Science and Higher Education for Women-Coimbatore" under Avinashilingam University and Masters Degree in Computer and Communication Engineering from "Kongu Engineering College – Perundurai" under Anna University. Currently she is working as an Assistant Professor in the department of Computer Science Engineering in Kongu Engineering College, Perundurai. She is a life member of CSI. Her area of interest includes Wireless Networks and Network Security. She has attended Seminars and Workshops organized by various Engineering colleges. She has published papers in International and National Journals and also in International and National Conferences.

INDEX

A

access, 69, 70
access state, 73, 74
active route, 113, 154
ad hoc, 33, 36, 51, 56, 72, 83, 96, 106, 109, 111, 112, 114, 115, 125, 126, 128, 130, 131, 132, 133, 142, 150, 151, 152, 153, 157, 160, 161, 163, 165, 166, 167, 171, 177, 178, 180, 198
Ad Hoc On-Demand Distance Vector (AODV), 107, 108, 112, 113, 114, 115, 128, 142, 158, 159, 160, 163, 166, 167, 178
aggregation, 19, 24, 121, 188, 193
Ant-colony-based Routing Algorithm (ARA), 108
Associativity-Based Routing (ABR), 108, 118, 119, 123, 153
asynchronous, 50, 86, 100, 102, 181, 187, 188, 190

B

bandwidth efficiency, 46, 83, 86
beacon, 16, 37, 96, 97, 109, 118, 119, 123, 135, 171, 172
broadcast, 2, 3, 13, 16, 17, 47, 66, 67, 68, 69, 75, 86, 100, 106, 114, 118, 120, 133, 134, 148, 149, 154, 156, 157, 163, 167, 171, 172, 185, 188
busy tone, 60, 61, 62, 63, 72, 176, 177, 178

C

cell resizing, 25, 44
channel, 1, 2, 7, 10, 11, 12, 16, 17, 24, 25, 29, 30, 31, 33, 35, 37, 43, 45, 46, 47, 48, 49, 50, 51, 52, 53, 54, 56, 57, 58, 59, 60, 61, 62, 63, 66, 68, 69, 70, 71, 73, 74, 75, 81, 82, 83, 85, 90, 94, 95, 96, 97, 98, 99, 100, 102, 103, 106, 124, 125, 134, 169, 171, 173, 174, 175, 176, 177, 182, 185, 189, 190, 191, 196
Clear-To-Send (CTS) packet, 52, 53, 54, 55, 56, 57, 58, 59, 60, 62, 63, 64, 65, 66, 68, 69, 70, 71, 74, 81, 82, 83, 84, 86, 87, 88, 91, 94, 97, 101, 104, 105, 106, 167, 168, 169
communication, 1, 2, 3, 4, 5, 7, 13, 15, 20, 21, 22, 24, 27, 28, 34, 35, 38, 39, 41, 42, 44, 46, 58, 69, 98, 112, 122, 137, 142, 147, 150, 151, 152, 156, 160, 170, 171, 173, 174, 177, 179, 180, 182, 183, 187, 189, 191, 194, 198, 199
concepts and architectures, 1
congestion control, 35, 36
connectivity driven, 183
contention based protocols, 45

D

data compression, 193
data driven approaches, 179
data fusion, 18, 19, 24
data packets, 35, 60, 62, 63, 64, 65, 67, 69, 71, 87, 113, 114, 130, 145, 148, 150, 155, 156, 157, 164, 177, 186
data prediction, 193, 194, 196
design challenges, 1
destination, 21, 32, 35, 39, 43, 53, 56, 63, 67, 70, 76, 77, 78, 83, 86, 92, 96, 97, 108, 109, 110, 111, 112, 113, 114, 115, 116, 117, 119, 120, 121, 122, 123, 124, 125, 129, 130, 131, 132, 134, 135, 136, 140, 142, 144, 145, 147, 148, 150, 151, 152, 153, 154, 155, 156, 157, 159, 160, 165, 166, 167, 171, 177, 184, 185, 191, 198
diffraction, 6, 9, 43
distance vector, 83, 86, 112, 128, 144, 163, 166
Distributed packet reservation multiple access protocol (DBTMA Protocol), 177
Doppler shift, 11, 43
duty cycling, 179, 181, 182, 190, 196

E

electromagnetic, 2, 3, 4
end-to-end delivery, 35
energy, 6, 10, 15, 17, 18, 19, 23, 26, 28, 32, 37, 39, 40, 160, 179, 180, 181, 182, 183, 184, 185, 186, 187, 188, 189, 190, 192, 193, 194, 195, 196, 197, 198, 199
energy conservation schemes, 179, 195, 197
energy efficient data acquisition, 195

F

fading, 1, 8, 9, 43, 173, 174, 175, 177
forwarding node, 138

H

Hop, 32, 69, 70, 89, 110, 113, 142, 178
Hop reservation multiple access protocol (HRMA), 69, 70
hybrid protocols, 192

I

IEEE 802.11, 45, 84, 97, 98, 99, 100, 101, 102, 167
information frame, 78
in-network processing, 179, 181, 193
interference, 1, 2, 6, 9, 10, 32, 36, 39, 43, 67, 100, 104, 157, 168, 170, 178, 192
inter-frame spacing, 101
introduction, vii
invalid route, 114
isolated, 78, 79
issues, 22, 29, 42, 45, 133, 192

L

layered architecture, 15, 16, 17
least overhead routing approach (LORA), 120, 121, 122
lifetime, 40, 74, 160, 180, 182, 184, 185, 197
link cost table, 116
link state, 120, 126, 131, 132, 139, 140, 141, 142, 144, 148, 149, 150, 161, 163
Load Balanced Routing Protocols (LBR), 143, 153
Location Aided Routing Protocol (LAR), 108, 143, 165, 166, 167
location driven, 183
low preference channel, 95, 96

M

MAC, v, 16, 17, 29, 30, 31, 32, 45, 46, 47, 48, 49, 56, 64, 66, 80, 81, 84, 87, 95, 98, 99, 100, 102, 104, 138, 140, 154, 157,

167, 168, 169, 171, 181, 187, 189, 190, 191, 192
MAC protocols, 49, 167, 189, 190, 191
MACA with piggy-backed reservation, 80
MACAW Protocol, 52
MANETs, 1, 12, 110, 142, 160
message retransmission list, 116
mobility of nodes, 12, 32, 33, 34, 35, 37, 43, 48, 99, 133, 134, 196
multicasting, 34
multichannel MAC protocol, 69, 95

N

networks, 3, 14, 20, 21, 22, 23, 24, 25, 26, 29, 30, 33, 34, 38, 41, 42, 44, 45, 48, 56, 72, 80, 83, 87, 90, 106, 108, 109, 111, 112, 115, 118, 122, 125, 126, 127, 128, 129, 130, 132, 133, 137, 140, 141, 142, 145, 146, 150, 151, 152, 153, 157, 161, 162, 163, 167, 171, 177, 178, 180, 198
non-isolated, 78, 79
Nyquist's theorem, 11, 12

O

on-demand, 72, 108, 109, 110, 122, 124, 130, 154, 155, 166, 187
originating node, 114, 115
Optimized link state routing protocol (OLSR), 125, 142, 161, 178

P

packet reservation multiple accesses (PRMA), 56, 57, 58, 106
path loss, 1, 6, 7, 8, 37, 43
polling, 30
Power, 23, 24, 32, 40, 41, 118, 119, 143, 160, 190
Power Aware Routing Protocol (PAR), 160
Preferred Link-based Routing (PLBR), 133, 134, 135, 136, 137, 143, 170, 171, 172
priority index, 93
proactive protocols, 112
processing, 11, 13, 15, 24, 25, 27, 28, 29, 32, 60, 108, 133, 139, 143, 144, 165, 178, 179, 193
propagation mechanism, 9
protocols, v, 16, 35, 45, 49, 50, 51, 56, 60, 64, 73, 87, 98, 106, 107, 108, 109, 110, 112, 122, 124, 128, 129, 133, 140, 141, 142, 143, 144, 145, 146, 150, 153, 158, 160, 163, 165, 166, 168, 171, 181, 183, 187, 188, 189, 190, 191, 199

Q

quality of service, 87, 129, 139, 142
query, 118, 146

R

radio, 2, 3, 4, 5, 6, 7, 9, 28, 40, 43, 46, 47, 57, 64, 69, 72, 90, 99, 122, 133, 134, 140, 158, 171, 173, 175, 177, 181, 182, 184, 186, 187, 189, 195, 198, 199
reactive protocols, 127, 137, 152, 162
real-time medium access control protocol, 83
reflection, 2, 6, 9
relay nodes, 36
repository, 28
reservation frame, 78
reservation mechanism, 49, 51, 83
route requests, 133, 146, 166
routing, 12, 14, 20, 21, 25, 32, 33, 34, 35, 37, 39, 66, 80, 83, 84, 86, 107, 108, 109, 110, 111, 112, 113, 114, 115, 116, 117, 118, 120, 121, 122, 124, 125, 126, 127, 128, 129, 130, 131, 132, 133, 134, 137, 139, 140, 141, 142, 143, 144, 145, 146, 147, 148, 150, 151, 152, 153, 154, 155, 157, 159, 160, 161, 163, 164, 165, 166, 167, 171, 172, 177, 178, 182, 183, 184, 185, 186, 197, 199
routing flags, 113

routing protocol, 12, 20, 32, 33, 34, 35, 37, 80, 83, 84, 86, 107, 108, 109, 110, 115, 120, 122, 124, 125, 126, 128, 130, 131, 132, 133, 134, 139, 140, 141, 142, 143, 144, 145, 146, 147, 148, 150, 151, 152, 155, 159, 160, 161, 163, 165, 167, 172, 178, 184, 186, 199

S

scattering, 9
scheduled rendezvous, 187, 188, 191
scheduling mechanism, 49, 90
scheduling table, 87, 88, 89
sensing, 15, 22, 24, 25, 27, 52, 56, 58, 59, 100, 105, 127, 162, 177, 179, 181, 182, 183, 192, 195, 196
sensor networks, 1, 15, 16, 18, 22, 23, 24, 25, 27, 28, 37, 44, 127, 162, 183, 191, 197, 198
sequence number, 110, 111, 112, 122, 130, 135, 138, 140, 141, 142, 145, 153, 166
service support, 101
shared broadcast channel, 46
sleep/wake up protocols, 187, 188
soft reservation, 72, 73
Source Tree Adaptive Routing Protocol (STAR), 120, 121, 122, 142
spectrum, 4, 5, 10, 21, 32, 45, 46, 69, 99, 100, 173, 174, 175
static, 73, 110, 141, 196
synchronization, 30, 47, 49, 50, 54, 69, 70, 72, 83, 86, 188, 192

T

throughput, 30, 31, 35, 36, 37, 45, 47, 66, 98, 102, 104, 140
Time Division Multiple Access (TDMA), 17, 19, 30, 56, 72, 76, 181, 189, 190, 191, 192
topology control protocols, 183
transmission, 4, 5, 6, 7, 9, 10, 11, 13, 15, 17, 18, 19, 21, 25, 30, 31, 32, 35, 40, 43, 45, 46, 47, 49, 51, 52, 53, 54, 55, 56, 58, 59, 60, 61, 62, 63, 64, 65, 66, 67, 68, 69, 71, 72, 73, 74, 77, 78, 79, 80, 81, 82, 83, 84, 85, 87, 89, 90, 91, 93, 94, 95, 96, 97, 99,100, 102, 103, 104, 105, 106, 109, 116, 133, 145, 155, 157, 159, 164, 170, 175, 176, 177, 178, 179, 180, 181, 185, 191, 192

U

uniform laxity budgets, 92

W

wireless, vii, 1, 2, 3, 7, 10, 12, 14, 15, 20, 21, 22, 23, 24, 25, 33, 35, 37, 38, 39, 41, 42, 43, 45, 46, 47, 48, 49, 50, 51, 56, 60, 72, 80, 83, 87, 90, 99, 100, 102, 104, 106, 108, 109, 112, 115, 125, 126, 127, 128, 130, 131, 132, 133, 140, 142, 151, 161, 162, 163, 169, 173, 174, 175, 177, 178, 179, 180, 181, 189, 190, 192, 193, 198, 199
Wireless Routing Protocol (WRP), 115
wireless sensor networks, 180, 181, 189, 190, 192, 193, 198

Z

zone routing protocol, 108